I0464545

1

THE WHIMS OF THE AGES

THE MOON THE MOTHER OF ALL THINGS

The Day of Doom and the Flight of the Gods

By J.M. Woolsey

4

DEDICATED
to the
Folk Lore Society of New York
and the
Folk Lore Society of London.

CHAPTER I.

THE MOON.

SUPERSTITIONS are chronic complaints, deep rooted and die hard.

The moon to the ancient world never ceased to be the land of mystery from which our religious whims and witchcraft have been drawn. Everything which occurred, every sight and sound was an omen for evil or good from the crowing of the cock to the muttering of fire upon the hearth, and all the antics of sun and moon were imitated and called divine mysteries, and all the trees, plants and fruits which had any likeness or affinity were dedicated to the moon for their virtues and healing properties.

Astrology arose from the belief of a connection between the heavenly bodies and the life of man and the influence of the stars on human destiny, and in the fourteenth and fifteenth centuries the astrologers were still in high repute. And all magic names, words, formulas, talismans and amulets but represent the talisman of the new moon, the great magician, cheat and trickster.

Lunacy in ancient belief was due to the influence of the moon.

The moon formerly regulated the growth of hair, the fullness of shell fish, fruit, meat, and caused steeples and pyramids to incline from their perpendicular.

Good tales spread rapidly and it is hard to find one like "Cinderella" or the faithful dog "Galert" or "William Tell" which has not been scattered from India to the Atlantic. They are better than true.

Most of our tales are of ancient lineage; the common heirloom and inheritance of mankind, with later adaptations of modern times which have wide distribution and though apparently strange and impossible are still true to their hidden prototype.

Twenty-eight years had passed the 28 days of the moon month and it was the last moon of the summer, and Jack (the moon) being a tinker was a spendthrift and could never save a penny: that is, the moon no matter how economical, he is obliged to part with a ring or coin every day and at the end of the month has given up his last coin, and his wife can never keep the wool she stores for it is pulled out of the bag, piece by piece, until the last piece of wool or light is gone, and the moon bag is dark and empty. The bag in which the woman puts scraps of sun wool is the moon bag.

The moon was known to govern the tides and the ebb and flow in obedience to the pulsation of the sea. Sunlight and darkness and the blood flow, pumped by the heart even as the tides are pumped by the moon, the great heart of the sky.

The ancient armies were accompanied by augurs and prophets who interpreted the signs, omens and portents of the sky.

Men sowed and reaped and bought and sold; they went and came and lived and died under the guidance and direction of astrologers, wise men and omen readers. They entered into all the details of their existence, they have permeated every industry and amusement of human life. They have outlived thrones and empires, survived reformations, overturned science and defied even common sense.

We still put fresh meat in salt brine three days to draw out the blood as the blood is drawn out of the moon for three days at every conjunction. For this reason children are not allowed to sleep with old people.

As virtue was supposed to be drawn from them even as the old black moon or witch draws the fire or life out of the bright moon as a vampire.

It originated the trial by ordeal when the person suspected of murder was required to touch the corpse when, if guilty, blood would flow from the lips of the corpse—the scene occurs at the end of every lunar month when the moon is slain and remains dead and dark for three days when the sun who was the murderer is obliged to touch the corpse and a stream of blood (the ring of the new moon) betrays the criminal.

The hair could not be cut, the nails pared or a tooth pulled without encountering omens and ritual.

The hair and nails were cut at new moon; they grew better in new light, like vegetables. Marriages were performed at new moon and to move in a new house at new moon, she will increase your store.

Girls were weaned at the wane to give them a lean slim figure and boys at full to make them robust.

Hogs were killed only on the increase of the moon for if killed at any other time the pork would shrink in the pot.

The British Druids consecrated their gold by dropping it in a lake as the sun dropped his gold in the moon lake at night. They had a lake in the mountains sacred to the moon called Helenus, and in this they cast linen, cloth, fleeces, cheese, wax and bread at a feast of three days, the time which the sun tarries upon the dark moon at the begining of every month. In heathen belief everything which passed through the fire would be preserved from evil, and they ran round their fields with flaming torches to bless and fertilize their land.

And the bell now tolled at death is the same bell of old and still rung in Asiatic countries to drive away evil spirits from the dying and dead man's soul.

Up in the courtyard of the moon, that land of tragic change and unrest, it was there all nations and creeds looked for omens. The farmer planted, sowed and reaped by lunar observations; the journeyman departed under her auspicious signs and watched to observe the new moon over the right shoulder. That moon that governed the dew, the rain and the vintage—that land of freak and disguise.

For ten thousand years people looked at that smoky chimney "Old Santa Claus" used to climb up and down there with Christmas gifts.

The ancients spent their lives watching the flight of birds, consulting oracles, omens, and observing the stars which had a supposed influence over human life.

Christians use these symbols as cross, dove, fish, anchor, horseshoe and rod, and all of them are the new moon ring and are moon symbols.

The moon was the sacred Mirror and ancestral shrine.

The moon woman feels she cannot live without her Lord, the Sun, wherever he goes. She feels herself imitating all his movements—at one time burned to ashes and again restored to life on the third day.

Pregnant women were not allowed to eat a hare or to look at a hare for fear their child would be marked with a hare or have a hare lip. This superstition originated in the moon, for there is the image of a hare in the moon and this moon hare was a witch wife and master of spells, transformations and magic.

It was the dead man's bone which prophesied from the tomb of Merlin—it is the dead horse head alone that is wise over the gate. That hidden hand, the dead man's hand, shall rise up from the grave as a magic wand to heal the nations.

The ancient mysteries were the celebration of the visible terrestrial and celestial mysteries as taught by the heaven and earth for every step of progress the luminaries advanced upon the great celestial highway was imaged and reflected upon the earth which followed the course of the sun and moon around the year which brought the annual return of their festivals, rites and ceremonies.

All our tales of raising the dead, cleansing the leper, healing the sick, working miracles, enchantment, transformation, transmigration, transfiguration, temptation, sin, fall, regeneration and all our rites originated in the visible changes and phenomena of the sun and moon and earth.

The broken ring the lover leaves with the maiden, the broken bone placed over the door, the one sandal left on the shore of the moon is the half ring, the sun lover who will complete the ring and the two welded around the moon as the ring of the new moon the wedding ring of springtime.

Isis, the Egyptian moon goddess, was the first to teach the Egyptians how to bleach linen. Isis is the moon; she bleaches her linen white every month.

Servian women will not wash a shirt at new moon; their linen, they say, would get "mooned" in the water—pucker and tear, but the stains and spots disappear when washed in the old of the moon. (Grimm, p. 715.)

This is why the laundress puts blueing in her clothes to make them white, as the moon first blues her garments and then bleaches.

This is why the wool and yarn of our grandmothers was dyed blue—the color of the blue moon—and then the bundles of rolls were pinned together with a thorn, which is the pin of the first new moon.

The moon was the snap dragon that bit off fingers when pointed at her. This was a wide superstition. It arose from seeing the little finger of the sun bitten off and stored in the mouth of the black moon as the new moon ring. Sometimes it is a nose bitten off, as in the Mother Goosey Rhymes; the moon maiden folding up her linen has her nose bitten off—"in came a black bird and snapped off her nose."

The discus or quoit is the disc of the sun and moon.

Twelve ounces to the pound in Roman weight to agree with the months of the solar year, the aisle of a church from ala "wing"; two wings, a wing on each side of the nave (navis), a rigged ship or bird; the two wings or forks of the new moon.

Ra and Sol in the eight notes of the gamut are the two names of sun deities, one Egyptian and the other classic.

They are our twelve panes of glass or windows of the three panes one way and four the other, and six in each sash for the two seasons of Summer and Winter.

We still wear charms upon our watch chain. The chain is the moon chain; the watch is the diminutive moon, the time keeper.

The king had an ivory throne, his palanquin was of ivory because the sun god sat upon the ivory throne of the moon and his festivals were held at conjunction of sun and moon.

The Hebrews sacrificed a red heifer to Jehovah and the Egyptians sacrificed red cattle to the good bull, the God Apis, and when Apis died they buried a red steer by his side for the same reason that red cattle were sacred to Typhon-satan or the evil principle; for Typhon himself was red, and this raised the strong aversion to red hair in human beings, particularly in the British Islands, not because they were Danes but red was Typhonian.

Ancient coins were stamped with moon symbols as an olive branch or owl, and coin means wedge or new moon, and money means moon.

Egyptians and Celts used ring money, still preserved in antiquarian museums. The Hebrew shekel was a silver piece about the weight of half a dollar. It had on one side a pot of manna, on the other a three-stemmed flower, or Aaron's rod. The pot of manna represented the moon, the bread giver, and the three-stemmed flower, the triad or three formed moon, always born on the third night of darkness.

The moon signs and telegrams originated clairvoyance, sending spirit messages and magnetic healing.

CHAPTER II.

SUN AND MOON IN RELIGION AND MAGIC.

THE sun to the desert shepherds was an evil destroying god that dried the fountains and burned vegetation, while the moon, the water god, was the nourisher of the vegetable world with the dews of night. The moon taught magic art; spirits that leave no footmarks but roam the air.

The moon was to them the silver door, and the heavenly door; the sun entered that door.

Religious ceremonies are all of them magic, as practiced to-day in our churches. Charms are worn throughout Europe against evil powers and spirits, fetishism and benedictions practiced under ignorant spiritual leaders.

Astronomy to the Babylonians was the mistress of all science.

A country filled with astrological quacks, soothsayers, necromancers, sorcerers and exorcists, to cast out demons and evil spirits, by a ritual of obscure and incomprehensible words and gestures, to move the Gods to compassion and break the spell or baleful curse. Their heavenly temple was the moon, and the Babylonians built a temple in its likeness and called it Babel, "the gate of God," which also gave name to its city of "Babylon," more anciently called "Seat of Life."

Before that mythical moon the diviner stood with Urim and Thummin repeating formulas and magical words endeavoring to avert, by magic, the decrees of fate.

We still stand in awe before the infinite depths of celestial space, reaching for the unknown.

The moon oracle is still in use as our almanac, which governs lunatics and is supposed to prevail over female periodicities and the weather, and we continually hear the inquiry, "When does the moon change?" It governed the tides and was supposed to shed dew (when no drop of water or dew ever fell from the moon.)

The new moon ring to the ancients was the seed or soul from which it grew—that soul or leaven which enlivens and gives life to the whole lump. The moon is short lived, and every month is burned upon a funeral pile to eliminate the soul and renew the lamp of life.

The ancients believed that all things had come from the moon, and that through the umbilical cord of the moon the earth drew its nutriment and sustenance.

Religions were made up from solar and lunar riddles translated from an age when everything is made to talk, and we hear the stories told from trees and stones, and the wind and the rain are made to talk and the flowers woo and wed, for the heavens and earth were reflected in that pool. There in that mirror the sun went to look at himself, there he sat down every evening when his day's work was done and the moon spread for him her silken pillows and crimson robes—it was the hotel where the tired old sun traveller put up over night.

Soma, the moon, was the sovereign of the vegetable world to the Hindu. All the ancient world was filled with signs, omens and presages; it was the chief lore; everything had its sign and director. In the building of a house, or the sowing of seeds, all were governed by signs, and if begun at an inauspicious time would fail.

That is the old lunar clock that ran twenty-eight days at a time and again wound herself up at the beginning of every lunar month.

Adam and Eve were married, there Cain slew his brother, there Christ was born, and there He died and arose from the dead.

MOON, THE GREAT REVELATOR.

Old Abraham dug his well there and planted a tree. Jacob slept there on a stone and then set the stone up for a memorial. It stands there yet, white and smooth, without a flaw.

In the depth of that dark moon is the cave of Machpelah or "double division," one part of light and the other of darkness, and the sword of light lies between them. That is the cave in which Adam and Eve hid themselves from the sight of Jehovah, the sun god. It is the sword or ring of light between the upper and lower world of the moon, the cave of the seven sleepers.

The new moon is the purse of gold, the box of ointment, the wish bone.

The new moon ring is the key of all things, the divining rod, the horn of plenty, the bread tray, the kneading trough. It is the whistle in the chest, the heart in the giant's body, the buried treasure, the hidden secret, the lost jewel and talisman, the bone of healing, and the pearl of great price. It is symbolism, in which nothing is called by its right name, but by borrowed figures. The new moon ring is the key of the Kingdom of Heaven (Mat. 16:19), key of death and hell (Rev. 1:18), key of the bottomless pit (20; 1 Rev. 9; 1 Rev. 20:1).

It is the collar of Harmonia, the girdle of Venus, the three mites of the widow in one coin, the first ring of the moon.

The new moon was the immortal ring. The one that fire cannot burn nor water drown, that gold purified by the fire, that arises from the conflagration of the moon on the third day. The Gulweig, the maid of the Norse tales, who was thrice cast in the fire and each time arose more beautiful.

The new moon was the signet ring, seal, and insignia of authority, and the evil eye and baleful look which withered life at winter time. He who looks through the loop of the wise man's arm can see spirits. And the wise man's arm is the **ring** of the new moon. Sometimes that new moon ring is said to be saved from the moon sea as a Moses or Jonah, that the sea has sworn a solemn oath to never drown.

There the sun saw not face to face but through a glass darkly. The sun was the artisan and image maker, "I and my father are one; he who hath seen me hath seen the father," for the moon was the image of the sun who was the father.

There the island was raised to receive the drowning Ulysses, the first land ever raised from the deep; from there came the wind and the storm and the thunderbolts of heaven.

From there came the signs, the signals and the alphabet of nature and the gods.

There were told the pictured stories and traditions of ancient times. From that mountain God's laws were thundered down; from that high court there was no appeal; there God's revelations were written with a pen of fire.

Little Cinderella, the sooty cinder girl, hid there the two dark nights of her wedding festival.

Peeping Tom of Coventry hid in there, and peeped through that crack of the moon wall at Lady Godiva, and that is the same Peeping Tom or Thomas, the infidel apostle, who doubted Christ until he thrust his hand through that same hole in the moon, the side of Christ, and we are just escaping from the wilderness of the old world ignorance and have overcome the despotism of religion with the freedom of thought.

There occurred Christ's wonders and miracles; there He

raised Lazarus. There the witch of Endor raised the prophet Samuel. There old Odin, the high god of the Norsemen, raised up the Vale the prophetess to learn the fate of Baldur.

There Wainamoinen raised up the dead giant to obtain the three lost words, the ring of the spring moon, the three rings in one. There the leper was healed, there on the third day Paul came to his understanding, there on the third day the memory came back to Peter.

Christ and Jonah were raised there on the third day.

That moon was the speech friend of the sun god, his right-hand man and counsellor.

Pliny says: "Hence we may certainly conjecture that the moon is the star of our life, this it is that replenishes the earth; when she approaches it she fills all bodies, and when she recedes she empties them; and from this cause shellfish grow with her increase and the blood of man is increased or diminished in proportion to the quantity of her light, and the leaves of vegetation feel her increase.

"Fluids are burned up by the sun. We have therefore regarded it as a masculine star burning up and absorbing everything."—Pliny B. 2, Ch. 102.

Pliny quoting Varro says the hair should be cut at full moon if we would avoid baldness. (Pliny B. 16, Ch. 75.)

If you open trenches by a waxing moon they will grow apart; but if opened by a waning moon they will grow together.

A wound inflicted in the dark of the moon was dangerous and malignant, if the sign were in the heart or feet.

The moon's phases were continually studied in gardening, grafting, gathering fruit, and cutting timber; as in the full moon plants were full of sap and in the wane timber must be cut and fruit gathered with least sap for better keeping.

Dew was erroneously believed to proceed from the moon,

and the growth of plants was referred to the moon goddess, rather than to the sun, and the moon presided over the birth of children, and a woman's time of childbirth was reckoned by moons and the periodicity of menstruation also was governed by the moon in their belief.

The ancients derived their conception of soul from the moon that life or vital spark which dies out and is yet not consumed but arises like Jonah and Christ upon the Third day. Again it was suggested by seeing the sun die at night and his soul seen to fly away upon the moon as in a dream of the night. Our "nightmare" originated from this scene of the moon. "Mare" is the name of the moon, the goddess of mare, "the sea." The hag and the strangler of the light and our heavy delirious sleep we call "nightmare."

In the creation of Adam God breathed into him the breath or soul of God and man became a living soul. With many ancient nations the seat of the soul was the liver, and again as with the Aztecs it lay in the heart, the soul of life. That ring of the new moon was the universal soul.

A saying in Cornwall when a child is born in the interval between an old moon and the first appearance of a new one, he will never live to grow up. Their saying was, "No moon, no man."

It is the soul or ghost of the sun or departed soul of the sun—even as the soul of the Egyptian Osiris the sun had to be confined in the black box of the moon, the winter tree, and his wife pursued on after him and obtained the bushy tree of the moon and cut off the outer covering and found a white pillar within which was the first ring or pillar of the moon, the one immortal which contained the soul of Osiris; it was the soul of Christ.

In mythology and metaphysics it is called thought, soul, mind, intellect, it is the first form Proteus, Protogonas, Prometheus the "Logos," the "Word" that became flesh, it defies age and the fiery furnace of the moon and rises from its ashes, it is the soul of the universe, it is the seed ring of the moon.

The American mound builders had the same superstition which prevailed among the Egyptians that the corpse in the tomb received visits from the soul which dwelt in its celestial abode. The idea of a double which belonged to the dead, a soul and body; and holes were bored in the skulls of the dead, that the soul might go in and out, and holes were left in the stone at the door of the prehistoric dolmens of Europe supposed for the same purpose as the apertures left in the most ancient of Egyptian pyramidal mounds for the communion of soul and body.

For the moonlight is the shadow or soul of the sun, it is the other self as the soul of the departed sun for from that moon the wise woman, the witch wife and the master of magic originated all rites and ceremonies, all whims and witchcraft, all wisdom and deceit that moon the mother of all things.

It originated cremation and the burning of the dead in imitation of the moon, the funeral pyre to eliminate the soul which arises on the third day from the ashes of the moon, and funeral obsequies were essential for the repose of the dead, as the soul of the dead moon was three days in rising. At the ancient Aryan funeral they walked three times round the spot with left side toward the corpse.

We keep our dead for three days and the funeral is held upon the third day.

The two principles of good and evil, life and death, creation and destruction are taught by the visible phenomena of heaven and earth around us. The wax and wane of the moon, her life and death—as a man will live until his fire goes out.

Magic necromancy, and wonder working through supernatural agencies, universally preceded religion such as rainmaking, disease making and sympathetic magic.

CHAPTER III.

THE MOON IN MANY ROLES.

T HAT moon was the ghost world where the disembodied sun was seen stalking in his night shroud as a wax image.

And the new moon, the original throne of the gods, was brought down by incantation. That moon was the incarnate wisdom of the universe, it taught revelations by dreams and oracles. This was old world belief.

Religion in every age has been purposely veiled in mystery; filled with ghosts, spells, magic, enchantment, sorcery, and dark sayings, where gods and men take beast form, and all inanimate things are gifted with speech and personality, and religions grew and changed with the growth and intellectual advance of the nation.

That moon is the smoking furnace which Abraham beheld as her fire had burned out on her funeral pyre. (Gen. 15:12, 17.) And the fiery furnace in which the three Hebrew children were cast.

That new moon is the tall pillar which Jacob set up and called it "the house of God, and gate of heaven."

It was in the gap of that moon, God the sun hid Moses while he passed by, and where Adam and Eve hid.

It was on the black wall of the moon, the bulletin board where Belshazzar saw the handwriting.

There Christ the sun god was transfigured at night. In that dark house of wisdom. There the ghost of Samuel was raised up for Saul for all the sun gods had to consult that oracle.

That moon is fairyland where the dead awake at night, where the Valkyrs raised to life the heroes slain in battle.

In the Roman religion caves were the abode of sibyls and nymphs.

In Greece caves were the temples of Zeus, Dionysius, Pan and Pluto; they represented the moon caves; they were the abode of oracles at Delphi, Corinth and Mount Cithaeron, and the worship of Mithras and the seven sleepers of Ephesus. Lot and his two daughters dwelt in that cave, and that was where Adam and Eve hid, and David hid there from Saul; the cave of the moon, the Macpelah for the west was the gateway of night through which the sun crept.

There are twelve houses on the Zodiac and the moon is set up in each one of the twelve. That house is the one pulled down by the strong man Samson, for in that winter moon the captive Samson the sun ground corn.

That House Christ was made to say He could pull down and rebuild in three days which is done now once a month, the old moon house disappears once a month and is reborn again on the third day.

It was the house of the seven gables, and the house forever divided against itself which shall surely fall.

That moon was the house of the Volsungs, in the northern legends the dwellings of kings when the world was young; its roofs were thatched with gold, and its doors were silver nailed, and its walls were hung with battle shields, and there was the throne of Volsung beneath the blossoming bower of the Branstock, the tree of Life "which sprang up in the midmost hall floor, and reared its blessings roofward."

Again it is the winter house of old farmer Celeus, where Demeter went to nurse the young sun prince, Demophoon.

It is the almshouse and the orphanage where old Abraham went down with Sarah, and the winter Pharaoh (Hades) took

Sarah away from him to stock his harem. The Holy Family went down there to winter every year.

Samson went down there in the winter moon to grind corn. Hercules went down there to weave on the loom of the harlot Omphale in that winter house. It was the treasure house and held the golden hoard of the sun. The Niblung treasure which was stored in the vault of the winter moon.

Hermes, the wise man, was the landlord. It was the celestial theatre where all the great cosmic tragedies have been played, for that moon stops in every one of the twelve constellations to play a piece appropriate for the season of the year.

In Spring it is the Beth-car, "house of the lamb" or Beth-peor, "house of opening," but at the end of summer it is the Beth-Nimrah, "house of Rebellion."

On that throne every New Year's day Ea, the Babylonian chief god, entered the "Holy of Holies" and seated himself above the Mercy Seat on the "Great White Throne."

God spoke to Moses from that Mercy Seat.

Jehovah (Yaveh) dwelt between the two cherubims on the floor of the Mercy Seat. "The Lord shall dwell by Benjamin and shall cover him all day long and dwell between his shoulders." Deut. 33:12.

And these shoulders are the two forks of the new moon.

To the ancient world the moon was animated by a spirit, she was mistress of the waters and controlled all animal and vegetable life.

That moon door was the mouth which opened and delivered oracles.

Soma, the moon of the Hindus, was the guardian of sacrifice and penance, he had twenty-seven wives, the twenty-seven lunar asterisms (days of the month.) He was "Lord of constellations" and "Lord of the lotus." He was healer of diseases and bestower of riches.

The moon was the great theater and celestial playhouse of the gods, no wedding or ball could be celebrated until the little Cinderella had arrived.

There was no spring until the arrival of that suitor whose footsteps were heard coming upon the mountains, "how beautiful are thy feet upon the mountains, oh fairest of ten thousand."

The moon with the Norseman was "the holy house of Odin, oh that hall of the "silver door." That the Goths and the gods have builded, to last forever more."

That was the house where Odin the sun stopped to drink and had to pawn his right eye the sun, and that was the pawn Gilgames gave and the one Judas had to give up to the temple treasure. Gilgames the hero sun of Babylonia stopped there and gave the tree of life as a pawn which was the ring of the new moon.

Jehovah, the Jew god, dwelt in the burning thorn tree of Mt. Horeb and spoke to Moses from that moon bush.

In the orphic verses souls come from the moon and return there as Manes.

And again the moon was the Machpelah, the "cave of double division," one of light and the other of darkness, with a sword between. It was fairyland.

The place of the "seven caves" and "seven sleepers," on one side is dark while the other is light.

In that moon house was the throne of God "where the Ancient of Days sat upon his throne whose garment was white as snow, and the hair of his head like pure wool."

He is the Norse God Odin and the gray old serpent the Ancient Saturn, the Ancient of Days (Dan. 7:9.)

The moon at the beginning chose that ring, the rich gift of wisdom which the sun when hidden behind the earth at night dropped in the pool of the moon waters.

The Egyptians tell us that Hercules rode in the sun and Hermes in the moon for the works of the moon seem to indicate wisdom and cunning.

And that seat was between the two arms of the new moon.

"And I will speak with thee from above the propitiatory from between the two cherubims" that are upon the ark (Seat of God) (Exodus 25:22; Ps. 80:1.)

They are the two arms of Moses held up by Aaron and Hur after the sun had set, which are the arms of the new moon in the west held up to prolong the light of day (Exodus 17:12.) The Christian pulpit to-day stands between the two forks of the moon altar which were the bull's horns.

The doors of the four cardinal points are the chief doors, being the entrance to the four chief houses of the sun.

Christ is made to say, "I am the door of life;" "Knock at the door and it shall be opened."

It was there Melchisedek the priest king met Abraham returning from the wars of Sodom; he was the first ring of the spring moon who became the high Priest of the moon altar that one without beginning or end of days.

That was where the black moon stone was rolled back from the sepulchre of Christ; that was the manger and birthplace of Christ and the cave where Zeus (Jupiter) was born.

There all the gods were born from the black cavern of the moon.

The child is not visible until the third evening in the west in the lap of the moon, the old black nurse.

All the miracles of Christ are still performed upon the moon by the magic wand of the new moon. Eden was in the east, for all nations of old turned to the east for peace and rest, toiling through the winter wars "until Shiloh come," going home to the east Cain went back there.

There in the moon were placed all the statues of the gods, that moon house was on the highway of the gods; it was the Travellers' Inn at one of these moon houses the sun stopped every month, where the sign of the Inn hung out at the cross-roads. In spring the sign of the ploughshare of the husband-man, in autumn the reaping hook and at winter the skull and crossbones.

For in the Hebrew account God made the sun and moon and hung them up for signs and wonders, and the seers divined by their lights and shadows.

But aside from being a house of sanctity it was used by the poets and mythographers as a celestial theater where gods and heroes played the tragedy of the year. The four greatest plays occurred at the four cardinal points when sun and moon meet at the equinoxes and solstices.

That moon house between the two forks of the moon was the "Great White Throne" on which Ea sat who was the chief god of the ancient Babylonians; it was the same seat of the Hebrew God who dwelt between the Cherubims. II Kings, 19:15.

There the Hindu God Vishnu slept on the couch between the two arms of Sesha, the serpent, again in the Norse it is the seat of Gripir, who sat in a chair of the sea beast's tooth. That is the great white throne and him that sat thereon (Rev. 20:11.)

It is the seat of the Norse god Odin between the two ravens, which are the forks of the moon.

It is the temple which Christ said he would destroy and rebuild in three days, which is done now at the end of every month.

And a representation of this moon temple was set up on the earth as an ark or sacred chest; a portable sanctuary containing the ten commandments, and on the ark was the Mercy Seat surrounded by the cherubim and the space between the cherubim was appointed as the meeting place between God and man.

Jehovah dwelt between the two cherubims on the floor of the Mercy Seat (Numbers 7:89), "and I will speak with thee from above the propitiatory from between the two cherubims that are upon the ark. (Seat of God, Exodus 25:22; Psalms 80:1.) Give ear, O shepherd of Israel, thou that dwellest between the Cherubims."

On New Year's Day, Ea, the chief god of ancient Babylonia, entered the Holy of Holies and seated himself above the same Mercy Seat "upon the great white throne."

That moon was the city of the gods and its earthly representative, the temple, had a silver door guarded by watchdogs, bulls, lions and protective animals or the cherubim or by two pillars set up like the two forks of the moon which guarded the sacred moon temple, sometimes birds, beasts or serpents instead of pillars.

That seat of Jehovah (Yahveh) was the same throne of Ea, the most ancient god of Babylonia, who sat upon the same "great white throne," the same seat occupied by Odin, the Scandinavian god, between the two ravens (horns of the moon) the same as that of Jehovah between the two horns of a bull which are the forks of the new moon when the Jehovah was represented as a bull, under bull worship.

Porphry says de, ant. Nymph.

The ancients had a notion that every person was born of the moon, before being born of his natural mother, and that the life germ invisible as thought permeated all substance as the sun rays were seen to come from afar and bring life; and as the dew (which to them came from the moon) was seen like Manna to bring food and life to plants.

That the vital spark was first germinated in the moon as the new moon child was seen to be born from the dark cavern of the moon through the silver door, and a like origin was postulated for man.

Through this same door the children came out of the moon ark of Noah, consequently the moon became the midwife and mistress of child-bearing.

Before this moon door sat Lucina (Juno) goddess of child-bearing as an old woman with her feet crossed and her fingers joined to retard the birth of Hercules, the sun child. She was the black moon who held the moon in darkness for the three nights, the utmost time allowed, for then the bright new moon ring child of the sun burst forth through the door of the moon, the Door of Life, but in the west in the after summer it was the Door of Death. Pandora opened that forbidden door, the same that Prosperine opened when she plucked the Narcissus, the sleepy plant which opened the way to winter Hades.

CHAPTER IV.

RELIGIOUS WHIMS.

THE burial ground had to be christened, it was unlucky even to walk over an unchristened grave, and the unchristened babe never died but wandered about the woods in solitude after apparent death and could not die.

The ancient philosophers and pilgrims travelled with staff and wallet—the staff is the stick or rod of new moon and the wallet the sack or pouch of the dark moon which was Benjamin's sack.

For our religion is but an elixir of moonshine.

The gods swore by the moon altar.

In the war of Jupiter with Cronos the gods swore by the cup of nectar, the moon cup or the cup of Soma, the moon of the Hindus.

The Shaman and medicine man carried a drum and rattle to imitate the thunder and lightning and a medicine bag for the black moon bag which moon was the great healer and medicine man, his ancestor progenitor and his wise moon teacher and all this from that far-off silver city with a language of speechless pantomime.

There to our fathers was the beginning of things, there was the shining "Door of Life," and there will be the judgment and the end.

It was in one part a garden of delight and in another the prison of the damned.

In early Babylonian times the moon worship prevailed over the sun. The moon was chief god and the sun was his vassal. The moon was the house where the gods met every month to transact business. That is where the old wise man Hermes, judge and counsellor, had his bench.

That was the altar of sacrifice where bulls were sacrificed to Jehovah; he, being a god, could not eat beef, but had the ox roasted, then he came around and breathed or inhaled the odor—"smelled" the sweet savor as he did when Noah gave him a reception at the end of the flood.

That is where the angels sang the world to harmony at the creation, that was the "Holy House of Odin, oh, that hall of the silver door, that the Goths and the Gods had builded to last forevermore!"

Adam and Eve were married at that altar. There the Cinderella danced at her wedding ball.

That is where the sisters obtained the patterns for their black and white robes and the priest his black and white gowns and the nuns their beads and the cross and the bell on the steeple to scare away evil spirits.

Pantomime was taught by the moon—dramatic representations without words through expression by attitudes and gestures, and the actors wore masks; it was a dumb show.

The moon was the revolving wheel of celestial activity. In ancient representations the sun as male is represented within the crescent of the moon.

The moon was the prison house of the soul which is seen to have a re-birth and escape from its mortal nature.

The Seraphim (Isaiah 6:2) are grouped around Jehovah's throne in his heavenly palace, their song shakes the foundations of the palace, they have human voices and hands.

For our religion has not been revealed from the Gods but has come up self taught from savagery.

For the moon measured time and set up mile posts on

the planetary track and as the moon governed the tides, time and tide became identified. We still speak of the tide of life and that life came in and went out with the tide and when a man lingered upon his death bed, they said "he is waiting for the tide to go out."

In Anglo Saxon and north countries the moon was masculine and the sun feminine.

Dutch, Dane, Norse, Saxon and Goth all have moon masculine, but in south countries the gender is reversed.

The moon revealed all the old religions; our lunar system of weather lore is based upon the ignorance of the dark ages; it is the sun which governs the weather and not the moon. The sun raises aqueous vapors and forms clouds; he is the author of every movement of the air and of every drop of water, dew, snow and hail.

Snake and demon worship was the oldest of all worship.

When men trusted in spells and charms, when demons controlled the elements and sorcerers exorcised evil and foretold the future, in that age they were subject to a soothsayer and men moved like a shadow only as directed by signs, as the cry of a beast or flight of a bird.

A merchant must lay the keel, launch and sail only by observing the omens. And the foundations of a building were laid with magic ceremony, when the omens were favorable, as at the rise of some planet or genial constellation, and the warrior waited for omens and auguries before engaging in battle.

For all religious rites and ceremonies and theological teachings have been evolved from natural phenomena and not from revelation or inspiration.

TEMPLE, CHURCH, AND SHRINE.

RELIGIOUS SUPERSTITION.

Holy water blessed by the priest is the imitation of the angel or sun priest at the moon fountain disturbing the water to make it healing for the leper that had no one to put him in. It was the fire stick, the first ring of the spring sun thrown in the bitter waters of Marah, the winter moon.

Fasting was taught by observation of the moon's wax and wane, and the unleavened bread of the dark moon for three days.

Circumcision. Genesis 17:11. "Ye shall circumcise the flesh of your foreskin and it shall be a token of the covenant between you and me—it is a representation of the scalp of the dark moon, showing the ring of the new moon or solar phallus, its great celestial prototype."

Cutting off the hair and castration were practiced in sympathy with the decline of the year. Cybele demanded the castration of all her priests in the temple and compelled them to become eunuchs.

The priest to-day pronounces that blessing to the blind congregation who have hidden their eyes—as the priest sun pronounces the blessing upon the dark moon.

Our temples still abound with the old imitation of nature, the columns for the tree, the altar, the nave and the aisle, nave from navis, the ship, and aisle, the wing that is the winged ship of the moon.

MOON, THE GREAT REVELATOR.
RELIGIOUS WHIMS AND LUNAR SUPERSTITIONS.

Ancient funeral rites were but lunar rites brought down to add to the multitude of human superstitions—unless the dead had funeral honors, their ghosts lay naked and wailed by the waters of Cocytus. The funeral honors were during the three dark nights of the moon when the deceased was wrapped in a pall, or sack-cloth and ashes, when the imprisoned soul escaped as the imprisoned soul of the moon is seen to escape on the third night as the ring of light or life.

Convents are moon houses and were often built on the banks of streams where nuns still live as shy spinsters who avoid men and ignore wedlock and nuptial life, who wear black robes with white bands, the more closely to imitate the moon mother, Mary, whose name means Mare, "the sea," and the word nun means a fish as the moon goddess, the Venus of old, was churned from the sea and had a fish pond by her temple filled with sacred fish.

Terah is a corruption of the word "Yerah" (moon).

Teraphim were moon images and oracles.

Laban said to Jacob "wherefore has thou stolen my gods?" Teraphim were consulted as oracles in divination similar to Roman Lares and Penates house gods.

Urim and Thummim were worn on the breast plate upon Aaron's sons, a divine oracle used in asking counsel of the Lord, sacred dice and images are found around the neck of the Egyptian mummies. The Babylonian Marduk tore the tablets of destiny from the breast of his foe and replaced them with his own.

Souls of the dead and house spirits, dwarfs and guardian deities, are still recognized by the Chinese.

CHARMS.

RELIGIOUS.

Religious mysteries are a representation of the mysteries of the sun and moon and natural phenomena.

The dividing curtain in the temple represented that dividing curtain of the moon, the light from the darkness, the sheep from the goats. Sacred bread offered to the gods, Christmas pie of goose in North England hog's head, hot cross buns, ancient Greek Bons, and Boun ox (Brand vol. 1, p. 155), a cake with two horns made of fine flour and honey.

Church bells were cast with ceremony and prayer and cast as a friendly aspect of planets. John Aubrey "Remains of Gentilism."

Healing herbs were gathered at new moon.

Anointing and circumcision were taught by the moon as the new moon was anointed by the sun.

From India to Europe children were passed through clefts of rocks and holes of trees called navels and when they came out were called regenerate; in imitation of the sun child passed through a hole in the black moon.

It is passing a child over the back and under the belly of a donkey to cure measles as practised in Donegal, Ireland. Church Folk Lore, Rev. I. E. Vaux, who adds that he was unable to guess the riddle. "It was taught by the moon which passes the young child of the new moon over its back and under its belly. It originated the saying, what goes over the

devil's back will go under the belly, for all healing and
miracles were performed upon the moon.   (Author.)

Sometimes a child was passed under the belly of an ass or
a piebald pony or nine times over the back and nine times under
the belly of a donkey. The donkey was the black moon, the
same ass which Christ rode. The same as carrying a patient
through the smoke of a lime kiln (new moon smoke) it was
healing. Going through fire to Molech.—Leviticus 18:21.

RELIGIOUS SUPERSTITIONS.

In all Christian countries in former times, it was customary to ring or toll a bell after a person had died while the soul was passing to heaven, to drive away evil spirits who stood about the house and at the bed's foot to seize the departing soul or molest it on its way.

It is still called "the passing bell" for the soul's passage, for the bell broke the enchantment and the bell was made to represent the black moon with a tongue inside and ringing. The bell would also heal the sick.

But a few centuries ago in England a man walked at a funeral ringing a bell to drive away evil spirits. (Northumberland Folk Lore.)

Catholic Christians still pray for the repose of the departed soul, for in the old creed, if burial rites were not administered the ghost would wander about to haunt the living.

These priests were the enemies of progress who kept the people in a spell of mystery under the pretended guidance of invisable powers.

The moon horn was very common in domestic use, and vessels were made to imitate the moon horn and we still retain the names in lantern (lanthorn) and (ink horn) for ink stand.

In the old superstitions we find a house was to be pulled down only at the wane of the moon.

The moon kept the sacred fire upon her hearth and was appointed to watch over and guard the family hearth.

But the gods were too far away, they must be brought down and dwell with us, live at our hearth and partake of our gifts.

We are still haunted with the shadow of the old faith.

Our religion is but heathen survivals of fortune tellers and sorcerers. Our gods are still anthropomorphic, the old man of the sky we still worship in human form, and old Jewish history is principally the feuds and wars of the sun and moon and the elements in the annual round of the seasons.

RELIGIOUS.

All the canonical and priestly vestments, robes and sur-
plices, scarf and scapulary, are but imitations of the moon's
changes, and patterns cut out by that magic pair of scissors,
the (new moon) which cuts garments out of the air, the
tonsure, cowl, hood, the apron, and all pontificals, pall, mitre,
tiara and triple crown were copied from the moon.

The Rabbis assert in the Talmudic traditions that the
vessels of the tabernacle were a fac-simile of those in heaven
and that a fiery ark, a fiery table and a fiery candlestick
descended from heaven for Moses to copy.

Ceremonial observances, sacraments, incantations, bap-
tism, christening, confirmation, Eucharist, consecration,
mass, extreme unction, transfiguration, transsubstantiation,
sackcloth and ashes, rosary and bead roll were all taught by
the moon.

In primitive Accadian and Babylonian times they relied
upon exorcisms and charms rather than medicine for curing
ills, for to them the earth and heaven swarmed with evil
spirits which produced the many diseases of human life.

And little statues with holy texts and sentences wrapped
around them were placed at entrances and doors.

And magic prescriptions were wrapped around parch-
ment or papyrus, as later the Philacteries of the Jews on
which were written portions of Jewish law. They were placed
on gates and doors and entrances as charms and exorcisms,
or bound around the hands and feet.

Each disease had a special evil spirit and this continued
on through early Christianity.

We still bring down celestial spirits by incantations and ritual into fetish objects as a tree or stone. We continue to bring them down into our temples to-day for the Moon is the Bethel or house of God; the same stone pillar which Jacob set up on earth, the image of the moon pillar.

For the principal religions and mysteries of the world have been taught by the moon, that wandering ghost of the night, the home of magic.

Our religious ceremonies are but the lunar antics of our heathen ancestors.

Our churches are adaptations of tree and ship and bird— a collection of ancient symbols.

The modern nun in black with a crucifix is that black moon with a rosary dangling from her hand.

From the wanderings of the moon, its phases and trans- formations originated a like emanation and wandering of the human soul.

All the principal religions, rites and ceremonies are but an astronomical allegory founded upon the life, death and resurrection of the year.

The sea digs down earth's walls and strews them on its floor. Earth groans and vomits fire; but again spring wakes to life what winter slays, and we build moon temples with spire and rod pointing upward in imitation of that fire rod of the moon and its altar.

And inside of our sanctuary we build the altar in a half circle to imitate the moon altar burned out and renewed every month, and inside of the altar we station a priest and ring bells which have a tongue inside to imitate the tongue of the new moon.

And would we destroy all this! Oh, no—ten thousand times, no; no.

Let the bells peal forth from temple towers—let Rachel
    mourn.
Let women weep and cast their veils to heaven—
"For mercy gives to charm the sense of woe
Ideal peace which truth can ne'er bestow."

CHAPTER V.

PLANT LORE.

THE new moon was the Fleur de Lis or "flower of light," the torch tree and lamp of heaven.

The moon to the ancients was bi-sexual author of magic and mother of all things.

Plants had to the ancient a divine origin. The divinities inhabited the trees and flowers, and these plants were sought out by the wise in plant lore, identified, and dedicated to the nymphs and deities: as Hyacinth had been changed to a flower of that name, and Daphne had become a laurel. It was supposed plants bore upon their face the image of their father and mother, the sun and moon, who were the great healers and destroyers—they were fire and water blended. The sun as fire and the moon as water in imitation of their heavenly prototypes.

In medical science each plant had a self manifestation and suggested its own healing powers by its form, color, and shape; each plant indicated its own office: as boneset was chosen because its leaves united and welded with the stalk, hence applied to fractured limbs.

As the healer sought for something pathognomic and symptomatic. And as the tides rose and fell with the moon, so life came in and went out with the tide.

So plants were supposed to fill and be fruitful on the increase of the moon, and fruits and meats, and all things on earth waxed and waned with the moon, and when the sun and moon waned, life withered.

In their belief nature had written out her wisdom on the trees and plants, and they studied the potency of herbs from the theory of similitude to learn their properties of bane and healing.

Plants were chosen to represent perpetual verdure or resistance to decay, or potent to counteract witchcraft—such charms as four-leafed clover or double-leafed ash.

The trefoil represented the trinity, and the amaranth or orpine (live forever) represented immortality.

The red rose was a charm against hemorrhage; nettle tea cured nettle rash.

Heart's-ease from having a heart on its leaf was a cure for the heart's disease, the blood root cured bloody flux.

Quaking grass cured fever and chills.

Swelling plants supposed good for tumors, long-lived plants contribute to the longevity of the patient, and sterile plants produced their like effects.

Salacious plants as mandrake promoted fecundity.

Saxifrage growing in the cracks of rocks was supposed to split rocks and called, as its name implies, the "rock splitter;" hence as outward signs denoted internal characteristics it was used to cure calculus.

From the observations of the sun and moon the air and earth arose, weather lore and agricultural lore, and plants were used as barometers and floral indices as they opened or shut or altered with the atmospheric changes.

The wishing rod and divining rod were cut from the hazel and holy thorn. It was cut from a wild hazel, the new growth of the present year. A forked branch cut by the right hand and by the moonlight of the crescent moon.

It must be a bough with a fork and twisted three times around itself; it must stand on the tree in a position that the sun could shine through, from east to west, or it was impotent. The finder bowed three times to the east, and made seven spells. Sometimes the fork of the fir or lime was used. All these observations and antics were in obedience to the moon. The forked branch represented the new moon crescent; it was twisted three times for the three revolutions of the moon; before the new moon appeared the finder bowed three times to the east for the same reason, for it required three trials before the crescent light could escape from the cavern of the moon.

Trees were split and held open by wedges, and diseased persons passed three times through, and after the tree had received the sin or curse it was plastered and bound up, and as the tree healed so did the patient.

The new moon was the Sesame of the Arabian nights, which opened the doors of caves and mountain caverns. The primrose in German legend is called key flower, because it appears at the time of the opening of the spring door of the moon.

The new moon was worn as a crescent, cross, sacred heart, horse shoe or scarabaeus, and the external resemblance of plants indicated their medicinal virtue in the cure of human ailments.

Protectives against sorcery and witchcraft by which cattle were struck to render them fertile and prolific, and to discover hidden springs and mines of wealth, they were cut with a forked end to represent the two forks of the new moon which was seen to bend over the moon waters.

Flowers were used in religious ceremonies and sacred rites. They possessed divine attributes and had a symbolic meaning and were woven into garlands which could be read by the initiated, forming a cross was healing like the new moon cross; or it was a medical theory from some external character of the plant. Yellow substances were used for the liver—red for the blood, and plants had their own virtues stamped upon them.

Plants were worn as charms and amulets. Doors and windows were decorated with them to drive off witches, such as St. John's Wort.

The Hindu hung up in the room a charm against witches, as the Scot and Briton hung up the rowan or mountain ash over the door. Men and women carried a stick of rowan in their pockets, and rowan berries were worn in the headband; likewise holly and bracken or heath for the same purpose, to avert evil, for bracken bore upon its root when cut horizontally the curve of the new moon or letter C for Christ. It is also the hoof mark of cattle. These talismanic plants had been engraved or painted in the likeness of their prototypes or stamped with the seal or image of the sun and moon.

The seed of the vipers Bugloss resembled the head of the viper, and its stem was spotted, supposed to cure the bite of spotted vipers, according to its natural signature and the law of similitudes.

Christians have chosen the passion flower from a fancied resemblance of its corona to the crown of thorns; other parts suggesting the nails of the cross.

Plants employed in sorcery were collected in certain stages of the moon, some gathered at the rising of the dog star, some from places where the sun never shone, some gathered for their Lethean or soporific qualities and placed in the witches' cauldron. It was the wizard's magic, and from these enchanters worked their spells.

Plants have memory and affection. The seed of the plant feels, thinks and awakes from the grave though long buried, and shoots up a stem the same its mother had, and remembers every tint of color and fragrance, as the human embryo in the womb of its mother remembers its former births and harks back through countless ages of evolution from the ancient sea, and during the nine months of its imprisoned life is seen to reflect the form of the fish and again reflect the hairy form of its animal progenitors.

The sensitive plants seemed possessed of personality, some vegetables, like the taro of the South Sea Islands, are seized with violent fits of trembling.

There is the weeping tree of the Canary Islands, a species of laurel which rains in the evening when the temperature is lower. The water comes out of the pores at the margin of the leaves and forms a little pond at the root of the tree.

CHAPTER VI.

COMIC AND WHIMSICAL TALES OF THE MOON.

THE moon in ancient story was a magic land, a land of marvels, and illusion, where a man is now a dwarf, then again a giant, or a vagrant travelling with a pack upon his back, or he is the King of "lone tower," or a man tall as a fir tree and has two torches in his head (the two forks of the moon).

In Celtic Romance, Joyce, p. 31, the names of the twelve sons of Balor the Mighty are "Large Heels," "Long Bodied," "Bare Knees," "Story Teller," and the wife of Balor was called "Crooked Teeth."

Other names as "Conan the Bald," "Luga of the long arms," "Nuada of the silver hand," like the classical Pelops the "burnt face," Oedipus the "swelled foot," Jason the man with but one sandal again called "Cold Foot," whose feet grew through the house and hung out in the cold at night. Again the moon is described as a man with long black hair, hooked nose and shoes turned up at the toes.

In Hindu he is called "shriveled foot," in Norse he is called Tyr the one armed man; again in Celtic traditions he is the tall bent lad with one leg or the man of the "Iron Mask."

And the moon children are the race who change their clothes and shed their teeth every month and had eyes of flame.

Again he is a man seen coming with his two ears sticking through his old hat. All these are the new moon; or a man seen without a head and his two legs spread out upon a barrel.

And a man having come to a fig tree and eaten until his nose grew a yard long, and again he came to a tree and ate until his nose was reduced to its normal size (at night he has the long nose which the sun cuts off by day), or a man buried in the ground all but his nose, and again a man whose nose was so long he couldn't hear himself sneeze.

The moon was the man of many vocations and Jack of all trades—he is at once a smith, a wheelwright, a tanner, a fuller and bleacher, a dyer, a weaver, a cobbler, rag picker, scavenger, water carrier, shoeblack, lamp-lighter and tailor.

He plays the part of a charcoal burner, gipsy, slave, negro, Moor, a hunchback and lives in Horn Castle. At another time he is herald, trumpeter, ambassador, and messenger of the gods.

The moon is a great actor and tragedian, story-teller, court-clown, jester, acrobat and magician. He is "Jack the Giant-Killer." He has a magic cloak, ring, table-cloth, rod, and boots.

He has an oil cruse which is never empty and a magic purse which is always full.

That moon house is one of eternal change, of war and strife, whose tenants have a fire every month and flee for their lives like Lot from Sodom.

He is that wandering ghost that now sits nude upon a dead body. Then enters and reanimates the corpse—that one who had carried off his wife's golden arm, and was haunted night and day with her cry, "Give me back my golden arm"—and on the third evening the golden arm of the new moon is brought back.

Again it is the little bit of dough which covered the whole griddle—the leaven that leavens the whole lump.

The moon was that royal coin upon which the sun stamps his image and superscription—he is the handsome stranger who can unlock her castle, be the night ever so dark; that moon which being struck with the white wand responds for good and gives good gifts, and again struck with the black or winter wand responds for evil and utters curses. She is that magic ball of yarn which unwinds until only a little thread of light is left and her husband, the sun, has the strength of twelve men (as "he" appears in twelve constellations).

The moon becomes a comic figure and a jester. Upon the stage in endless transformations he becomes a ring, a tree, rod, or serpent, and in his journey through the twelve constellations becomes a ram, bull, lion, scorpion, goat, fish, as well as man and woman, and at the end of the summer season all this menagerie is gathered into the moon, which is the winter ark of Noah or Deucalion, which is the farm house of old Celeus, the heaven father.

The moon was the great actor and tragedian of the sky. He invented time and made the first clock which would run for twenty-eight days. He could drop to sleep and shut the world in darkness—and again awake in a moment and flash light to the uttermost parts of the earth.

He played every part known to the human stage from King to the court-clown and slave; he could take all forms and transformations—he could make himself tall as a pyramid and again shrink to the size of a needle.

The dual nature of male and female prevails over the earth through the animal and vegetable world—sun and earth—day and night—summer and winter.

In the attempt to reveal man to himself and explore the divine revelations, ideal forces were called up by symbolism and the celestial spirit was brought down to the stone or tree image or fetish objects by incantations and ritual.

The moon was the author of magic science, and the ancient mysteries and the doctrine of sympathies between substance and shadow, and the visions in fairy-land or the ghost world.

It was second sight—seeing things in vision which could not be seen in the flesh. It was where the life of the moon as a dog or horse was bound up with that of his master, the sun.

That ring of light is the first principle of life which survives the conflagration of the moon and rises from its ashes and rends the grave or black vault of the moon.

That pillar ran down to Hades, for when Ishtar the goddess went down there she beheld that pillar or rod of life, the same which guided Aeneas through the mysteries of the underworld.

At the root of that pillar which ran down to Hades the Egyptian god was buried—Admetus drew up his wife Alcestis from that dark moon pit after the sun had sat below the western horizon.

Auguries and inauspicious days, shadows and echoes, Divination by Bible and key and shape shifting all taught by the moon.

The Hebrews divined by lots and Urim and Thummin, and by ephod which meant "yes" and "no"—or good or evil—"won or lost."

At the Feast of Tabernacles water was drawn from the well of Siloam and poured upon the altar "in order that the rains of the earth might descend to you." This was an old rain charm pronounced in many lands of dearth.

The spirits of the dead in belief could hurt those on earth, and at the Irish wake the living were in fear of the departed soul who might appear and injure the living, and for this reason they shaved their heads and tore their clothing to make themselves unrecognizable.

The Chaldaean houses were purified by driving from them a host of evil spirits.

Demons both material and immaterial belong to old world inhabitants of low culture, spirits of the sea, earth and sky, winds and devils of malignant type were propitiated by offerings. In ancient Babylonian demonology the demons controlled everything—every part of the human body. Even in Europe at the present day they believe in the corn spirit "Lyncanthropy" and the "Were Wolf." These demons were the cause of disease and death; they still exist in epilepsy and are expelled by spells, sacrifice and prayers.

The Norseman believed that a genius in animal form accompanied a man through life as a guardian, as the Manitou of the American Indian; that a man had his familiar, a tutelary spirit guardian or watcher, and they raised the souls of the dead, as the Witch of Endor raised the ghost of Samuel and as Odin rode to Hell to wake the wise Vale, the old prophetess, the winter moon.

As the phantom appeared to Odysseus of old Folk belief, or the Incubi or bloodsuckers in sleep the Banshee or death warning, wild Huntsmen Hounds, fairies of fairy knolls, the corn spirit which had to be reverenced as a Demeter, the cereal deity as a man or a woman, sometimes as a child (the corn with an animal embodiment) and receive sacrifice.

The black stone of the moon was called the "covering stone," and was over the pit of Hades. It was the black stone rolled off the sepulchre of Christ, and revealed His white linen robes, and the black stone Jacob rolled off the moon well for Rachel; that pit of the under world was

"Orcus," where dwell the images of deceased mortals, for the earth was a hollow dome through which the sun and moon passed and the scenery was copied from the moon, the same as the white river seem to girdle the moon was brought down to girdle the earth.

The moon was an alien house; the sun and solar race were strangers there and foreigners.

Though the sun every year redeemed their land in spring, yet in the after summer time their old evil nature returned and the sun was banished or put to death.

That moon house is where the Philistines hung up Saul's armor in the house of Ashtaroth (1 Sam. 31:10); it is the sun's armor hung up on the moon temple at night and winter. It was in that temple of Dagon Samson the sun captive ground corn in winter captivity until his strength returned in spring, when he pulled down the old Temple to the ground.

It is the house of Rahab the harlot, the last moon house of winter before entering the spring signs.

It is the house which Christ said He would destroy and rebuild in three days, for the moon is burnt out and arises from its ashes on the third day; the house forever divided against itself which shall surely fall.

"Thou that dwellest between the cherubims shine forth. Turn again, O Elohim, and cause thy face to shine, and we shall be saved."

It is the sun shining between the two arms of the moon. The name variously called Jehovah, Elohim, El Shaddai, Adonai or Javeh.

CHAPTER VII.

LUNAR PLAYS AND AMUSEMENTS.

PUBLIC SPORTS, AMUSEMENTS AND LUNAR PLAYS.

PUBLIC sports and amusements were arranged under the sanctity of religion as in running, horse racing, football; for in games they leaped for the slippers, they danced for the new moon ring, the Cinderella, the maid of the spring equinox.

In Blind Man's Buff the moon man was blinded and dressed in hairy attire. The play was called Blind Buck and David for it originated when Christmas occurred in the sign of Capricorn the goat; he is the hairy Esau and covered with a goat skin. The one hoodwinked is then turned around three times, like the blind moon, and questioned.

"How many horses has your father?" "Three—red, white and grey." (The three changes of dress or collars of the moon.)

"Turn about three times, catch whom you may."

Counting out rhymes were formerly priestly incantations by elimination. There is one left after all the rest have been counted out, and it is the new moon which comes on the stage, that one left of Lot's men, that one left to tell Odysseus the fate of his companions in the hands of Circe, the sorceress.

Jumping the broomstick from behind the door is the new moon leaping out of its confinement in the black moon castle, for the new moon was the broom that swept the dirt or the cloud off the moon.

Dominoes have 28 pieces, white and black, for the number of days and nights in a lunar month.

Cards, 52 in a pack and 4 suits of 13 each, making 52 cards, the number of weeks in a year.

WEDDING AND MARRIAGE CUSTOMS.

The old man who lived in the moon presided over marriage and tied the marriage knot with a silken cord,—the thread of the new moon which was also the cord cut at death.

In a Chinese belief, Yue-Laou, "the Old Man of the Moon," unites with a silken cord all predestined couples which cannot be annulled; that silken cord is the slender thread of the new moon which connects the two forks.

In Hindu marriages the bride makes seven steps or steps into seven circles and joins her hands in a hollow form while repeating ceremony. The seven steps represent the seven months of winter for the spring wedding occurs in the seventh month, they are the number of the seven devils that went out of Mary Magdalen, the winter months.

Again the maiden plighted troth by clasping hands through the hole of a tree or stone, the hole in the stone or tree represented the cleft in the black moon stone where sun and moon clasp hands.

The twin bone of a chicken or goose broken and then laid over the door represents the bone of the new moon broken, and the half ring put over the door before weddings—a lover's pledge.

Marriages were solemnized at the time of the new moon, and the bride celebrated her "honey-moon;" this was the honey-moon of Samson at midsummer in Leo when he married the harlot and took her into a new house for the winter season.

Samson, the sun, had reached his highest ascension at

mid-summer in the constellation or the house of Leo, "The Lion;" the end of the summer months here the sun chariot is always turned back toward the south, and here the winter harlot Delilah betrayed him to the winter Philistines.

So closely are all our rites and ceremonies woven of lunar fabric that our marriage word "connubial" means to veil. In modern Egypt the bride is kept veiled and cannot be seen by her husband until after marriage. Our women still wear lunar veils and the time for weddings was fixed at a plane- tary conjunction in India—the meeting of sun and moon. They still marry in some countries without having seen their betrothed according to the old lunar myth the wife has not seen her husband nor ever met, like Cupid and Psyche and told she had a beast husband, which is the Jacob who sleeps the first night with the dark moon, the sore eyed girl as Urvasi in Vedic who was never allowed to see Pururavas naked or without his garments, which is at night in sleep with his garments removed. The sun comes into the cham- ber of the moon but cannot see her for three nights; as soon as the naked sun beholds the moon she will vanish.

Under the old Spartan rule the bridegroom was compelled to visit his wife in the dark, suggested by the sun who visits the moon, his wife's chamber in the dark.

In Layard's "Nineveh and Babylon" a marriage is described where the bride was led to the bridegroom's house, and kept in the corner of a dark room for three days during the festival before the bridegroom was allowed to approach her.

When wine was drunk at Jewish weddings the glass was broken and the bridegroom walked three times round the bride and took her by the right hand even as the sun every month marches around the moon for the three dark nights before he will take her by the hand.

WITCHCRAFT.

Witchcraft was born in prehistoric ages and was universal with the human race. Witches existed among all nations, savage and civilized and was strong in Greece and Rome and was a superstition of the old world and the wide world. It prevails to-day among savage and half civilized nations.

All over Europe witches were daily executed. They raised storms and changed themselves to cats, hares, and wild beasts—rode on broomsticks, pokers, goats and dogs.

Witchcraft was a supernatural power erroneously obtained by entering into a compact with the devil by which they abjured God and Christ and renounced the sacraments of the Church.

A witch of the early times was the dispenser of spells and was in league with evil spirits.

We quote from the most reliable statistics that in England during a period of two hundred years, over thirty thousand were executed for witchcraft and even this was outdone in other countries.

In Geneva, Switzerland, five hundred "witches" were burned in three months of the year of 1515.

In Scotland from 1560 to 1600, the number of victims for the forty years was 8,000, and this was fully equalled in France, Spain and Austria under papal bulls, as well as Protestant decisions, down to the time they were repealed by parliament in England and Scotland in the early part of the Eighteenth Century.

From the time of the promulgation of the bull of Pope Innocent VIII against sorcery in 1484 until 1782, 300,000 women perished on the imaginary charge of witchcraft. ("Encyclopedia Americana.")

It cost one hundred thousand lives in Germany alone between 1450 and 1650, and the old Puritan, Matthew Hopkins, went through the country testing the people by piercing them with pins and needles under a commission of Parliament, and in the County of Suffolk alone sixty persons were hanged for witchcraft in a single year. Bells were marked with inscriptions to give them power to drive away devils and lightning, and science was denounced and pulpit preachers made desperate effort to defeat science.

Ben Franklin was branded as "the arch infidel" for drawing lightning from the clouds, and preachers rained anathemas upon him from the pulpit for putting up lightning rods to prevent God from striking with thunderbolts.

Bells were rung to keep off witches and hail storms and days of prayer were appointed to end drought, which was a companion to witch hunting, under the old diabolical statute of Jehovah, the sun god, "Thou shalt not suffer a witch to live" when no such thing as a witch ever had an existence; that shows how much the all-wise Jehovah knew.

And every rite and ceremony in the Christian Church to-day is whim and witchcraft. Although we have made great reforms and have banished Original Sin, Satan, hell, torment and most of the Old Testament.

ZODIAC.

The Zodiac contains twelve constellations and has the ecliptic or annual path of the sun for its central line.

It is divided into twelve signs, each thirty degrees in length.

Previous to this, in the older times, the Lunar Zodiac was older than the solar which indicated the monthly course of the moon around the earth, and was divided into twenty-seven or twenty-eight Lunar Mansions, in each of which the moon dwelt one day of the month, for in the older time the months were calculated by the moon which went around the earth and sky in about four weeks or twenty-eight days, and the months were calculated by the revolutions of the moon.

The signs in their order begin at the Vernal Equinox or the place where the sun crossed the Equator when coming north in Spring and began at Aries, the Ram, but the older Lunar Zodiac was older than the solar, and was divided into twenty-seven or twenty-eight Lunar Mansions, in each of which the moon was represented as dwelling during one day of the month.

Invocation to the moon from the Babylonian, "O Moon, chief of the gods, king of the gods of heaven and earth, of the stars upon stars, which dwell in heaven great—into this temple when joyfully thou dost enter the holy buildings and the temple of 'The Great Tree.' "—From the "Records of the Past," vol. 5, p. 147.

The moon was the original temple and dwelling of the gods.

The moon pillar, that nightmare of the ancient world,

built the Pyramids and founded the doctrine of immortality and regeneration having been brought down to earth by incantation.

But old orthodox religions no matter how sacred or divinely inspired are fast giving away, and old fossil creeds are yielding to the continual growth and higher advance of religious thought.

It was the wisdom of the moon, that magic looking glass that reveals the movements and positions of the sun as an index and keeps the divine records of the solar family.

Our religion has come down from the ancient world filled with devil worshippers practicing mysterious rites—oriental priests and magicians dealing and communing with ghosts and disembodied spirits, that was the throne of God, the house of the silver door and the altar of sacrifice.

There the little sun child (ring of the new moon) was born, the Moses in the reeds and rushes of that moon marsh, for the moon was the head fountain and source of waters to all the ancient world.

The land of phantoms, of witches and ghosts' spells and enchantments, the land of giants and dwarfs, the hall of the divine masquerade, there occurred the transfiguration of Christ, the ghostland where the absent Sun as a spectral illusion was seen stalking upon the moon wrapped in his winding sheet at night.

There upon the moon the dead awake and the disembodied spirits appear in fairyland. The moon was the stage of mythology, and the home of the gods. All the great tragedies of the year were played upon the moon.

The yearly tragedy of Adam and Eve and the serpent is still played upon the moon and earth every year.

Sun and moon both furnished the scenery and decorations of the lunar stage. The parties are principally called by their badges and allegorical names, not from totemism but a high order of learned symbolism adopted by the priesthood.

Simple ideas were expanded and characters were multiplied to fill the stage, the one moon maiden became ten thousand to fill the Harem of Indra, the horse became a herd of horses, the seven became seven hundred, the ten became ten thousand.

Behind this dark curtain of the moon there are artisans painting scenery and landscape, there are sculptors of ivory and carvers of precious stones.

In that darkened chamber the fairies are weaving bright raiment for the Cinderella who will win the prince at the ball.

The curtain rises and falls, there are changes of dress and position, servants, cup bearers, slaves and bond women wait upon the guests.

Jewels are brought from foreign lands and the Jacob arrives with costly gems and the necklace of pearls. The banquet is provided with electric lamps, vases, salvers, spoons, knives, and rings, all of which are manufactured from the one ring of the moon—the magic table silently offers its gifts of all good things for the rich banquet, for the new moon of the spring equinox is the gospel of glad tidings, a phantom to which the great sun king gave a soul.

It is nature's tragedy where sun and moon play a monthly role, when they are in conjunction and cross each other's track. It is then transferred to earth, which is in constant sympathy with all their movements. When they fail or retrograde, the other, too, wanes, and when they are in their plenitude of power the earth puts forth vegetation and all nature rejoices, for the earth and moon and sun were in constant sympathy.

The moon has many names in the Hindu, called "maker

of night," "marked like a hare," "lord of the constellations," "having cool rays," "white rayed," "marked like a deer," "lord of the lotus," "drawn by white horses" (names translated from the Hindu).

Again she is called "sunlit hill," land of rings, and called the black castle and the iron castle.

The ring of the new moon had endless names—it was the hammer of Vulcan and Tubal Cain and the hammer of the Norse god Thor, the ox-goad of Shamgar and the flail of Gideon—it was the pen of Moses, the sword of God and the trump of the Archangel, the rod of creation, the plow of Noah the farmer, and of Cain the tiller, and of Elisha, who ploughed in the field with twelve yokes of oxen. It was the necklace of Harmenia and the girdle of Venus. It was the branding iron Xerxes used to brand the Hellespont—the fetters he let down in the sea to bind it and the whip with which he scourged the waters.

It was the golden cup from which he poured the libation and the golden bowl which he threw in the Hellespont for a bribe.

The moon was a magic lantern; it is now a sunlit plain or a castle with towers, again a green, grassy lawn, or a gloomy forest, again a sea with a ship tossed on its waves— again a sea of milk or a sea of blood.

It is the dove or quail, the prophet of spring, or the raven which forbodes the winter, or a wishing-stone, tree, cap, or a pair of scissors which cut all manner of garments out of the air, or a table cloth which provides everything the appetite may crave and all wishes gratified.

LATE TELESCOPIC REVELATIONS OF THE MOON.

From the most remote times the moon was the most attractive object of the sky. From its many changes of face, its connection with the ebb and flow of the ocean, and the supposed author of the dews and influence on human life and vegetation, it became the home of witchcraft and superstition.

The moon is nearly spherical with a diameter of 2,163 miles, a little more than a quarter of that of the earth, moving at an average distance from the earth 239,000 miles, its surface one thirteenth, and its volume one forty-ninth that of the earth. We can only see a little more than half of the moon or about three-fifths of the moon. It has no light of its own, and its interval from sunrise to sunrise, is a month; it has a long sunshine of two weeks, then a night of two weeks; the temperature at the surface, even at the hottest, probably never rises above the freezing point of water (Professor Langley) and in the long night the temperature must fall to something like 200 degrees below zero. It has no air, no water, no clouds, no vegetation.

The moon has rugged mountains from fifteen to twenty-five thousand feet high and crater holes twenty miles across and three or four miles deep with no air, no water, no clouds, and has the fearful cold of empty space hundreds of degrees below zero.

We are 240,000 miles away from the moon.

To a man standing on the moon the heavens above him

are totally dark. Owing to the absence of air there is no dif-
fused light in the lunar sky, consequently the heavens are
as black as ink in full daylight.

—Garrett P. Serviss.

The moon shows great contraction, and has had great
tidal disturbance caused by the earth, and she has great fis-
sures, caused by cooling, and has absorbed all the air and
water from the outside. Craters appear in almost every direction, circular rings with diameters from half a mile to 100
miles, for the moon is a burned-out world which has
absorbed what little air or water it may have had.

The great illusion! Instead of angel choirs, a land that
has no sound, instead of a holy hill of Zion and archangels we
find a land of baleful shadows and ghostly spectres.

Was ever a sight like this! A scene beyond words.

A Godless, soulless land of despair.

Gone is our moonshine religion. We have been worship-
ping ghosts for ages and the pillar that guided Israel through
the wilderness is but a streak of moonshine and God's throne
but a desert of dread.

The telescope has been our revelator and not St. John of
Patmos nor Moses.

It has revealed to us not a garden of the gods but a
mountain strewn with volcanic desolation that has no air, no
water, not even a living thing.

The moon is like myriads of other dark planets, lifeless
and scattered through the universe—dark suns, supposed to
be more numerous than luminous stars.

The universe swarms with dead moons and lifeless worlds;
whose lights have gone out from the mindless, soulless universe.

Billions of inhabited planets revolve around fixed stars, continually gathering an accumulation of cosmical dust, for motion had no beginning and space no limit.

We are but cosmical dust—the solar system infinite and life universal.

We have a man-made religion, the worship of the occult forces of Good and Evil, mysterious rituals, dark and esoteric, handed down from the ancient highways of Asia, prophetic dreams and visions, communications with the Spirit-World and mysteries of Future Life.

But old gods and religions have vanished and our present cannibal religion is already in its last throes.

Our earth has an orbital speed of $18\frac{1}{2}$ miles per second as it travels around the sun (this is around the sun), but it is rushing together with the sun at the rate of about twelve miles per second toward a point in the northern sky, but the telescope had been improved and laid bare the caverns of the moon, and there arose a great mumur among the priesthood, "who is he that has betrayed the gods," and it was found to be Galileo, and he was summoned and brought before the inquisition and there arose a great murmur among the people, saying, "nay, this is Galileo, one of the most noted scholars in all Europe, and the idol of the people," but the priesthood cried out, "curse on his wisdom, he has betrayed the gods:" for at that time chemistry was considered a black art, and men of advanced ideas were confined in the madhouse.

The works of Copernicus were considered heretical and the man who discovered the mariners' compass was accused of witchcraft.

Galileo was forced by the inquisition to abjure the Copernican Theory and was detained in the palace of the inquisition and condemned as a heretic to incarceration and to recite once a week for three years the seven penitential psalms and became hopelessly blind.

Science had at last outgeneraled religion, for the telescope had laid bare the throne, and the hiding place of the gods, who fled to the unknown realms of the ether as they did in the time of the Babylonian flood.

PART II.

THE SIGNIFICANCE OF NUMBERS

INTRODUCTION.

THE SIGNIFICANCE OF NUMBERS.

THE Orientals had great reverence for numbers and ascribed to them almost unlimited powers. Numbers belonged to the mysteries. The Ancients divined and prophesied by numbers as a part of religion. Numbers had rhythm, harmony and hidden properties. Numbers like letters to the ancients possessed mysterious virtues.

Numbers were magical and prophetic by which fate and destiny were revealed, and a great many significations were attached to numbers as divinity, unity and intellect.

Numbers were made to agree with preconceived ideas; when once a number was established, as one predominating in nature, all things were pressed into its service.

When seven became a divine number Thebes must have seven gates and Jacob must serve seven years and the nation must have seven wise men.

In Egyptian belief Thoth (the moon) was the inventor of numbers. He measured time and was the scribe of the gods, and the number of gods and Cabiri varied according to the number of powers in which the cosmogonic energy was divided.

Numbers among the Chinese had gender, the odd numbers were masculine, including one, and the even numbers feminine.

The Abury temples contain according to Higgin's Anacalypsis the cycle numbers 650-608-600-60-40-30-19-12.

Many numbers were sacred and some divine, the way they reproduced themselves seemed like magic.

The sun, moon and stars moved by numbers and reappeared at their place of beginnng after a certain number of days, months and years for numbers, letters and speech were the gift of the gods.

According to the conceptions of Pythagoras, numbers represented the properties and essence of all things, and served as elements in the construction of the universe.

Odd numbers were preferred by the ancients; in Japan to-day houses are set one front and the next back of that line from the street to make the line uneven, and in height the same care to make one high and the next low, for there is luck in the odd numbers, and the Chinese have their streets zigzag.

Pliny says, "Why is it that we believe that odd numbers are the most effectual and that there is luck in odd numbers?"

With the Chinese, odd numbers belonged to heaven and even numbers to the earth.

Five and ten among early races were the most noted and perfect numbers, for it was by the fingers that men first learned to count, and used in counting and numbers still called digits or fingers, and they measured by hands and fingers.

The Esquimaux could only count to ten. The Australians could count only up to four, the Brazilian Indians up to three, and up to ten was the limit and stood there. After that it was many fives and many tens of fingers.

With the ancients, numbers were an irresistible power, for they represented the elements and the gods and powers of nature and became associated and identified with them.

The Romans measured by tens and hundreds.

The Babylonian god Merodach cries to his heavenly father Hea for assistance in casting out an evil spirit from the afflicted one. "O my father, the baneful charm like an evil demon acts against the man."

Hea speaks, "A number make, this man is unwitting: by means of the number he enslaves thee."

Hea says to his son Merodach, "The number thou knowest not; the number let me fix for thee."

Numbers were as efficient as words.—"Ancient Babylonian Charms." Records of the Past, vol. third.

Among all ancient nations the planets ruled the affairs of man.

The Egyptians divided the human body into thirty-six parts, and each part was under the protection and government of one of the decans or aerial demons who presided over the triple divisions of the twelve signs.

There was no profane science. Medicine, physics, and astronomy were all priestly and religious, for letters and numbers were all revealed either by a natural or spiritual revelation.

Ten in the Pythagorean system is said to be the most perfect of all numbers comprehending in itself all difference of numbers, all reasons, species and proportions of numbers.

NUMBER TWO.

Two belongs to the oriental dualism of good and evil, day and night, sun and moon.

In the Dualistic theories, the conflict of the two divinities, good and evil, as two brothers, are common to civilized nations in general.

THE DIVINE THREE.

THE TRINITY.

In every world a triad shines forth of which a monad is the ruling principle. "Ancient Mysteries of Thomas Taylor."

The Universe was a three-fold world, one above the other.

The trimurti as creation, preservation, and destruction was the union of one god in trinity.

And that new moon contains the three hidden rings in one, the trinity visible on the third night of the moon's darkness.

The three-fold trinity of the Norsemen Odin, Vile, and Ve; Odin, Hoener and Loder; Odin, Thor and Balder.

The equilateral triangle was one of the symbols of the Greek Bacchus and Hindu Siva; it has three sides.

The Hindus wrote the name of their supreme power in three letters, A U M, correspnoding to his three dominions containing the trinity of beginning, middle and end, and our "I am" is a corruption of this, and our G O D contains these three significant letters, so does the Hebrew Y O D.

Ewald has shown that the Phoenician Trinity corresponds to the old Accadian, and that these three gods were but names of the sky, earth and underworld.

Three is heard in trio, triumph, trumpet, throne, trident and three toothed, for the first ring of the moon is made of three rings, three in one, for it is not visible until the third night; three nights in making, and the Lord passed by Elijah, first as a whirlwind, then as an earthquake, and finally as a still small voice, and that voice is the first ring of the new moon, appearing on the black moon on the third evening of her darkness.—I Kings, 19:11.

There were three grand divisions of the physical universe, sometimes the heaven, the earth and underworld, or three elements, air, earth and water.

THIRD DAY WHIMS.

The fortune teller turns her teacup upside down and three times round as the new moon or revelation of light appears, on the third day of her darkness. Sometimes the Bible is used for divining, instead of the cup, while repeating, "Turn, Bible, turn."

The throwing of three handfuls of dust in the grave repeating, "Earth to earth," "dust to dust," "ashes to ashes," was an ancient Egyptian custom, for the dead moon arises on the third day, so did Christ, and the Christians, it has been said, cannot call a single rite or ordinance their own.

Women walked three times round a church in order to have children, for our church and temple represent the moon; the new moon ring is born on the third day.

As in Sutteeism of the Hindus burning the dead was moon mimic, believing that the soul went out with the flickering flame, even as the moon is burned to be raised to life on the third day.

Again imitated by keeping watch three nights over the dead for the three dark nights of the moon.

It was by a pilgrimage to the different stations and holy meeting places and fasting three days that a blessing was obtained.

Candles are kept burning beside a corpse three nights to frighten away evil beings.

And treasures were revealed by dreams, but must dream the same dream three nights in succession and not tell anyone. If any one is told they will find the money all turned to charcoal, that is, if the dream is extended for three dark nights of the moon she will reveal the gold which is the new moon purse.

Pythia, the priestess of Apollo, at Delphi, sat upon a tripod or stool of three feet.

As the light had to be born again in the moon, so must man be reborn, and every child was said to be first born of the moon and then flashed to earth as by an electric wave on the third night.

That is why Jacob was obliged to sleep three nights with Leah, the sore-eyed woman (dark moon), before he could enjoy Rachel, the bright new moon.

The marriage ceremony of all nations was a moon festival, and the passage of the sun and moon upon a wedding tour to a new constellation in the three dark nights of the moon. In a Polish custom, the bride walks three times around the fire then sits down and washes her feet.

The wife in primitive times had to be chosen from a stranger race or clan.

Making a love bridge represents the white ring of the new moon of spring thrown across the dark moon water. Sometimes it represents the ford of the moon waters or the bridge over which the spring sun crosses at the spring equinox, on the third evening of the moon's darkness.

In Egyptian marriage customs the husband cannot see his wife's face until the wedding is three days old, and again a man divorces his wife by simply repeating "I divorce you" three times—that is, "I divorce you, I divorce you, I divorce you," and in this same way the bans or public proclamations are repeated three times in the Christian church.

The three classes of Gods in Vedic of the heavens, air and earth.

The Trimurti is a modern conception of Brahma, Vishnu, Siva.

The old triad of Babylonia was Na, Ea and Mulge, which became under the great sacerdotal reform and new official religion, the union of Accadian and Semitic, the trinity of Anu, Ea and Bel, and these were all children of Zicu, or Zicara, "the sky."

There were three beings of the lower world, Hermes was the three headed or tricephalus, and Cerberus, the infernal dog, was three headed, and most of the gods and goddesses were triform. Zens was triform, Diana was triform, but gods could take any form in mystic story of three smiths, the three gods in one, the trinity and triad, tripod, trefoil, tripartite cross and bull's head, are the common property of the ancient nations.

We see what three means when the three brothers, Jupiter, Neptune and Pluto, divide heaven, the sea, and underworld between them. Jupiter had the dominion of the air. A three-fold world seems well defined among the Egyptians as Nut the sky, Seb the earth, and Amenti or Hades the underworld. There are triads of father, mother and son, triads of heaven, earth and lower world, triads of beginning, middle and end, triads of creation, preservation, destruction.

Three is the most common for the three worlds or elements, as Brahma, Vishnu, Siva; Jupiter, Neptune, Pluto; Odin, Henir, Loki; Jehovah, Christ, Satan; Horus, Ra, Tum; Heaven, Earth and Underworld.

And this three fold nature was expressed by a triangle, thus $\wedge$; it represented also the three parts of time as am, was, been or Jehovah, the being who was and is and is to come, and three became divine. Yod represented the idea of God and Jovah (I am) that is God exists, which was too holy to speak; for which Adonai was substituted in its place, that is Adon or lord.

At the Passover they ate three unleavened cakes. Abraham at his tent saw three angels and Sarah prepared cakes of three measures of meal for the angels (Genesis 18:6); correspondingly there were three natures to man.

The Hindu God Brahma said, "I am the beginning, middle and end."

The mysteries were celebrated for three days, servants bowed three times to the King.

The trinity of Hindu gods was Brahma, Vishnu and Siva, the three persons of the Hindu trinity spirit, matter and time, the Babylonian Anu, Ea and Bel.

Egyptian, Horus, Ra and Tum.

Christian, Father, Son and Holy Ghost.

Osiris, Isis and the child Horus were the great divine triad of Egypt from which our Joseph, Mary and Christ are modeled.

The Hindu priests assume the sacred cord of three threads, the mystic symbol.

But the great Hindu triad of Brahma, Vishnu and Siva is of late date. The old Vedic gods of the first rank were Indra and Agni and after these Varuna, Mitra and Surya.

In the early trial of Babylonia were the three Gods, Anu, Bel and Ea, who represented respectively the heaven, the earth and the sea.

Noah had three sons, Shem, Ham, Japheth. Xisuthrus had three sons, Zerovanos, Titan and Tapetos.

The Hindu Manu had three sons, Sama, Cama, and Pra-Japati (Faber's "Origin of Pagan Idolatry.")

The German Mannus, son of the god Tuisco, had three sons who became the ancestors of the Germans.

Polyphemus had by Galatea, three sons, and Saturn had three, Jupiter, Neptune and Pluto.

Mercury was called Triceps, Bacchus Triambus, Diana Triform.

Prosperine had three heads, the Trimurti was the tri-mighty of the Saxons, the trimegas of the Greeks and the termagnus of Latins.—Celtic Druids. Chap. V., Sec. 19, Higgins.

Under the worship of Fo in China they have the triple god.

The triple God is all over Siberia and in the Imperial collection at Petersburg is the figure of a triple god seated upon a lotus, and in all these they remain one in essence, one in three and three in one.

A being triform, according to Faber, was worshipped by all the ancient nations of the earth.

God worshipped under a triple form in India as Brahma, Vishnu and Siva.

In Persia Ormuzd, Mithra and Ahriman, they are creator, preserver and destroyer.

Roman, Jupiter, Neptune and Pluto.

The Persians believed the origin of the gods was from the three elements fire, earth and water, hence three natures in deity and man.

In the Norse system there were three gods, Odin, the father of the slain, Loki, the destroyer, and Haenir, who wrought the hope of man at the beginning of the world, the redeemer.

The first trinity (Asgard, Gods, p. 56) is Odin, Vile and Ve, which correspond to air, water and fire. They were sons of Bor.

Three was the basis of Druidical worship, sacred officers were divided into priests, prophets and bards, the three classes in Hebrew as High Priests, Priests and Levites.

Dodona, the most ancient shrine of the Greeks, produced oracles by the vibrations of a caldron struck with a whip, the lash of which consisted of three chains. The caldron represented the dark moon and the lash is the sun rod of the new moon ring, which is throughout mythology the triad or three in one.

At the fire festival on St. John's Eve, held upon the hills, men, women, children and cattle were driven or leaped through the fire.

As in the Norse tale of the maiden called Gulwig, who was passed three times through the fire and every time appeared more beautiful like the thrice born of the mysteries.

Hecate was a triple goddess and the sphinx triple bodied. They were different forms of the same conception.

At the Passover they ate three unleavened cakes.

In the ternary classification of the ancient Druids of Britain who abounded with triads of which they had three hundred, they enumerate:

The three first settlers called the pillars or civilizers, three benevolent tribes, three usurping tribes, three awful events, three chief casters' works, three great regulators, three controllers, benefactors and primary sages.

Three masters of mystery and so forth.

Three immense stones were set up in the stone circles of Britain to represent the cell of life.

The thrice repeated oath of the Hindu was sacred and inviolable and if not fulfilled, vengeance was repeated upon the offender.

Enchanters use three leaved plants, such as enchanters'

night shade, which is three leaved and its parts arranged in threes.

Arabia and Gaul were divided each in three parts.

There are three colors on our flag.

Jonah and Christ were three days in the pit of the dark moon.

Noah had three sons, because there were three worlds or three seasons.

An old Arabian legend says Jospeh was in the pit three days, which was true; the moon pit in which he was cast is dark every month for three days, for he is the first ring of light upon the moon.

Persians believed in three elements, fire, earth and water.

Christ has three trials.

The cock crew three times for Peter, and there were three kings at Christ's birth, the three seasons.

Athena was the daughter of Neptune and the Tritonian marsh, which means the new moon born on the shore of the moon sea or in the trough of the new moon ring.

The three original seasons were spring, summer and winter, the autumn was added afterward.

Three among all the ancients was the Chief of Magic numbers for it contained within itself a beginning, middle and end. A trinity of three persons in one god.

Neptune had a trident and the thunderbolt of Jove was three forked.

Siva, the Hindu god, had a trident three pronged.

There were three visible divisions of the universe, heaven, earth and mid-air. This was perhaps the origin of the sacred number three. In its multiplications it always reproduces itself, hence the number became a symbol of eternity.

It is a number you cannot destroy:

$$2 \times 9 = 18 \qquad\qquad 6 \times 9 = 54$$
$$3 \times 9 = 27 \qquad\qquad 7 \times 9 = 63$$
$$4 \times 9 = 36 \qquad\qquad 8 \times 9 = 72$$
$$5 \times 9 = 45 \qquad\qquad 9 \times 9 = 81$$

adding the two figures of each product you again obtain nine, and each of the products is a common multiple of three and each one of the products divisible by nine.

Three was the most divine and sacred of all numbers, it was the number of the trinity of gods. Mind Columbus set out with three ships from Spain on his voyage to discover America and from Porto Rico again Ponce de Leon set sail in three brigantines to discover the fountain of perpetual youth.

The words of the Semitic languages (Hebrews, Aramaeans and Arabs) are all triliteral symbols. Every word has three letters and only three which constitutes the skeleton of the language. ("Ridpath Races.")

Among Greeks at the time of sacrifice the altar was sprinkled three times with a laurel branch using holy water.

The obolus was a three cent piece of silver placed under the tongue at death, the three represented the three rings of the new moon in one, which have been welding for the three dark nights and appear as the three cent piece to redeem the dead moon.

For creation, preservation and destruction were the three principles of the universe.

The triad A U M of Asia was the triad of the new moon, the three magic letters or the three magic words of old Wainamoinen of the Kalevala, the seed ring which Waina- moinen set in his moon ship. In the most ancient conception the three fates were one and called Urd but were afterward divided into a Trinity.—Asgard and Gods, 223.

Antigone sprinkled dust three times on the corpse of her brother's remains, which was equivalent to interment (three dark nights of the moon.).

In Arabia a guest is entertained for three days as a rule and a prescribed limit.

Neptune had a trident; even the dog, Cerberus of Hades, must have three heads, that dog is the three headed god of India and three gods with one head of the Christians, it is the one god in unity who guards the treasures of the year through the dark period of winter.

The three is expressed by a triangle. It is one of the sides of the pyramid.

It is expressed by a god of three eyes, and this is the reason we have three degrees of comparison and the third time conquers as the third shot of the sun wins upon the moon on her night of darkness.

At the door of Hades stood the three headed watchdog, Cerberus, and the Egyptians gave three cries at the grave of the dead, and Charon required an obolus or three cent piece of silver for passage money, and at an Episcopal burial the sexton gives three shovels of dirt in the grave, saying at each time, "Dust to dust, ashes to ashes, earth to earth."

We have the three shouts in court from the cryer "O yes" or "hear ye" three times when he announces the sitting of the next court, and three groans were given for contempt.

The third trial conquers and the third blast of the trumpet or beat of the drum, the thrice repeated injunction.

And all this because the moon sleeps three days, before it will reveal the hidden word or the buried treasures or heave up Jonah, the hidden ring.

For the universe was divided in three parts, heaven, earth and hell or the underworld, and the gods were three, the trinity.

They are the three mites of the widow cast in the black box of the moon.

That is old Pandora's box, the woman of all gifts.

The divining witch cup found in Benjamin's sack on the third day of the journey was the cup of the new moon and the priest repeats Amen three times at the end of his discourse.

This wisdom came at the third stroke of the moon rod, or at the third word or thrice repeated charm word; on the third day Jonah arose from the moon deep.

At the third day Christ arose from the same place.

The divine fasted and slept three nights in the moon temple to obtain a revelation.

It was the third dream of Pharoah or the third night sleep of the moon that brought the revelation to Nebuchadnezzar that hand writing or the golden letters upon the black wall of the moon, at the third night of her darkness.

It was at the third crow of the cock that memory came back to Peter.

Again the three days' journey so often repeated in the sacred writings.

Most of mythology was written under a year of three seasons, the time of sowing, of harvest, and of the inundation of the Egyptians, and in Babylonian three gods divided the universe, Anu, Bel and Ea.

Sometimes three stands for the three dark nights of the moon, again the three seasons, again the three elements or the three worlds.

Hel rode on a three legged horse.

Multiples of three, as three times three, seven times three, ten times three and three hundred continually occur in mythology.

The triangle in a circle represented the three seasons.

NUMBER FOUR.

The sacred four was marked out by the sun at the two equinoxes and solstices, and the points connected by lines into a square, and the sacred square is the ground included between the four cardinal points.

Among the red men the Winnebagoes place four sentinels at the four cardinal points, and the Scandinavians have four dwarfs.

The Hebrews had the four quarters of the heaven and earth represented by four animal figures, and have four letters for the name of their God, and the heavens propped up by four pillars.

We have a four armed cross and four weeks to the month.

Each tribe of the Germans was dvided into four classes, nobles, freemen, vassals and slaves.

In Ezekiel, Ch. 1, we have four living creatures, and every one had four faces and four wings.

The Hindu Brahma has four heads and Krishna has four arms.

The Oriental gradation of the four classes or castes is on the same scale of deterioration as the four ages from the golden age of spring to the iron age of winter.

The Aztec victim was marched four times around the temple before his heart was torn out.

NUMBER FIVE.

The letter A is an abbreviation of the Saxon Ane which meant one, the letter A in Hebrew meant one or unity and the letter A retains the form of the five stars—in Aries, which is the first sign of the Zodiac.

The Ace of Spades represents the black moon, the wise man, the leader of the pack, that ace is the thumb, the chief of the five fingers.

To the Nahuas, the fingers five were the five fates or five works or "five fields," for by the use of his fingers and hands, man worked out his destiny.

There were five primordial elements, earth, air, water, fire and ether. And the five colors and the five musical notes were created in reference to the same number with the Japanese.— "Griff's Religion of Japan."

There were five planets of the Babylonians,—Merodach, Ishtar, Ninib, Nergal, and Nebo,—corresponding to the Jupiter, Venus, Saturn, Mars and Mercury of the Romans.

There were five colors—blue, white, green, yellow and red—among Brahmans, and Brahmans have five sacred observances. Five is a favorite number among Buddhists.

There were five kinds of knowledge, for at that time there were but five months,—five for summer and five for winter.

The first numbers were the fingers, one finger raised as I, two fingers raised as II, three fingers III, four a's IIII, up to five, which was the four fingers crossed by the thumb which was contracted to V, and ten was a double X or two V's joined at the apex.

The first week known was of five days and represented by the thumb and four fingers, and after that the six day week.

The old lunar year at first consisted of ten months,—five summer and five winter. It seems that five and ten were used long before astronomical numbers as complete and always retained after the solar numbers six and twelve.

Five became the hand that moved the revolving year and had five fingers. It turned the celestial fire drill and was succeeded by the week of six days and then by seven, and it would seem that five had at one time represented the five moon kings of the ancients or half the year at a time when the year was divided among ten heroes, a number more ancient than the later division twelve months.

There were five telchines who were sons of the sea and powerful enchanters.

There were five planetary divinities exclusive of sun and moon, making seven.

Five stars—the "rainers"—stood in the face of the Bull in the constellation of Taurus.

In Astrology the universe was expressed as the four regions or the four quarters of the sky with the sun in the Zenith.

The Chinese have five capital virtues, humanity, justice, politeness, wisdom and rectitude. Their musical instrument called a kin has five strings.

In the Hindu philosophy the body consists of five elements to which it returns, and they were space, air, fire, water and earth.

The Egyptians enumerated five elements, and chief of the five was the celestial ether, which was Ammon the Egyptian Jupiter. It was the vital ether which quickened all animated nature.

The qualities of the sun and moon in Egyptian belief were five—fire, water, air, earth and a quickening spirit or agency, and these five objects were gods and given names. To the animating ether, Jupiter; to fire, Vulcan; to the earth, Demeter or Ceres; to water, Oceanus; and to the air, Minerva. (Diodorus, Book I.)

Cadmus destroyed all but five serpents in the building Thebes (he chose five) and David chose five stones to dethrone the winter giant Goliath. Five kids were slain at Sodom in spring before entering the spring equinox, or the month of renewal.

Abigail rode upon an ass with five damsels of hers that went after her when she rode to David and became his wife. It was the wedding of sun and moon in spring (1st Samuel 25:42).

Buddha had five disciples and there were five wise virgins and five foolish who took no oil.

The chief priest at the human sacrifice of the Aztecs had five attendants dressed in white who officiated at the altar

The Chinese left the land of the five summits and the four canals to settle in the far east.

The Olympic games were celebrated every fifth year, and the Olympic festival continued five days.

There were five lords of the winter Philistines and they gave to the Israelites as a trespass offering at the time they sent back the Ark the images of five golden mice and five golden emerods (1st Samuel 6:3).

The five Kings war against Gideon (Joshua 10), the Danites sent five men to seek out an inheritance, the five intercallary days. (Judges, ch. 18).

It is the king and five others, again it is the princess and five maids.

According to Plutarch, five wax candles were used at marriages.

Mt. Meru was surrounded by four other sacred mountains at the cardinal points or supporting mountains, which completed the sacred number five.

Japan had a national "fifth day," a day of rest, instead of one in seven.

The Hindu people had the "Pancha tantra" of five books, and the Hebrews have the "Pentateuch" containing five books.

A sacrifice to celestial gods was with an odd number, but sacrifice to terrestrial with an even number.

Plutarch's Numa—"Numas Precepts" in Chinese numerals, days with odd numbers are strong and the even are weak.

The cross of five dots is found as old as the bronze period of the Lake-Dwellers and on Etruscan tombs.

A picture was found in Herculaneum. It is of a marriage where a sorceress is divining with five stones.

Five was a complete number. Ireland was early divided by ancient geographers before the Christian era into five zones.

It will be found continued in the five "Hance towns" and the "five towns of Aragon" on the frontier of Aragon and Navarre, to the extreme of the Pyrennees. There were the five cities of the Plain and five Kings of Sodom.

Again there were five kings of the Amorites who went out to fight with Joshua (Josh. 10:5).

Five wounds were given to Christ.

The Koran commands five prayers five times a day.

The Hindu makes the five daily offerings.

Five crosses adorn the altar slab in churches,—the number of Christ's wounds.

Egyptians had a five-rayed star of horns.

And the sun cross as a six-rayed star was found with the Hittites.

The Buddhists, in Buddhistic lore, Mongolians, especially the Lamaites, have the five sacred in preference to the seven, as five zones and five planets.

Buddha was followed by five disciplines who deserted him.

Five husbands had that woman at Christ's well, and Draupadi the heroine of the Hindu epic of the Mahabharada married the five Pandu princes, who shared her in common.

With the Creative Council of five gods at Hermopolis, the "House of the Five," its temple became "the temple of the five,"—they are Thot and the four auxilliary gods of Egypt.

NUMBER SEVEN.

The seven days of the week were given to the seven planets, which were all that were known at that time,—that is, sun, moon and five planets. There were seven stars in the cluster of the Pleiades and in the Great Bear and seven colors to the rainbow.

The Bull (Taurus) originally opened the year he held the Pleiades, the seven stars of spring.

Throughout all nations the planets were assigned to the days of the week, or the days of the week were called after them the seven day cycle, (Higgins' Anacalypsis) and the same day allotted to the same planet universally over the world.

These planets became the Rishis of the Hindus and the Archangels of the Persians and Jews and the Archites of the Deluge.

The paths of the seven, that is, the five visible planets and the sun and moon making the seven rishis or great gods. All lie within this Zodiacal belt.

The seventh sign—the sign of spring, was healing in the spring, the seventh month.

The seventh well healed Naman the Syrian, and the seventh son was the healer.

The seven planets of the Assyrians and Persians had power over the elements and ruled the sea and the rivers.

The earth was at first divided into seven zones in stages one above another like the pyramid, but later it was divided into four houses, as in Egypt, corresponding to the cardinal points and like that of the Hebrew four.

Canaan was originally divided among seven nations.

England was divided into seven kingdoms under the rule of the Saxons called the Heptarchy.

Jacob served seven years for his wives (Genesis 7: 2).

Dan. 4:32.—Nebuchadnezzar driven forth to eat grass as an ox until seven times should pass over him, which are the seven months of winter.

In Chaldea seven times had the body of the sick man to be anointed with the purifying oil. By sevens had the knot to be tied by the witch, and seven times must the sick man be dipped in the water.

There were seven archangels of the Jews, and consequently seven Christian churches.

There were seven saints:

> St. George of England.
> St. Denis of France (Dionysius).
> St. James of Spain.
> St. Anthony of Italy.
> St. Andrew of Scotland.
> St. Patrick of Ireland.
> St. David of Wales.

There were seven wise masters.

Seven wonders.

Seven sciences of the Middle Ages, and they were Grammar, Logic, Rhetoric, Geometry, Astrology, Arithmetic and Music.

The Egyptians had seven mortal sins and sprinkled blood seven times upon the altar.

The Memnonium of Thebes uttered the seven mysterious vowels to the Egyptian priesthood.

In the Astrological and Cabalistic book of the Apocalypse, John is partial to seven and twelve. Seven is repeated twenty-four times and twelve is repeated fourteen times.

There were the seven magnificent Gods of the Babylonian pantheon.

And the seven-headed serpent of the Accadians "which lashed the waves of the sea," and the seven spirits "born without father or mother" in the encircling abyss of the ocean.

Clean beasts were taken in the ark by sevens and the ark rested in the seventh month (Gen. 8:4).

There were seven good kine and seven good ears in Pharaoh's dream.

There were seven gates to the entry of Hades, and at each one Ishtar is divested of ornaments and apparel until she arrives naked at the last gate.

There were seven altars of Balaam and we have seven days in our week only because they knew of but seven planets, and they gave a day to each.

At the seven gates of Thebes each had upon it the name of a planet.

The Egyptians divined by fashioning words from every third and seventh letter.

The number seven is the five miraculous loaves and the two small fishes.

Thebes of Boeotia was called Heptapylos from seven gates one within another.

Rome was built on seven hills.

The City of Ionia or the Syrian Antioch was built on seven hills.

And there were seven subterranean inclosures of the Kingdom of Queen Allat, the mistress of the Babylonian Hades.

The seventh day of the seventh month is one of the greatest festival days of Japan, it is the seventh of July (Japanese fairy world).

The ancient Egyptians walked the sacred cow seven times round the temple at the winter solstice (Dupuis).

There were the seven Sabbaths of years or forty-nine years.

There were seven stones consecrated to the seven planets.

There were seven doors to the cave of Mithras.

Pan had a pipe of seven flutes and Apollo a lyre of seven strings.

There were seven stories of the tower of Babylon surmounted by the eighth, which represented the sun.

There were the seven great spirits of the Persians, which became the seven Archangels of the Jews.

And the virtue will be in the touch of the seventh sign of spring in Taurus the Bull at the spring equinox, as Scorpio will be the seventh and evil touch at the autumnal equinox.

They are the seven ears of corn and kine in Pharaoh's time.

The book of fate was composed of seven books and the seven prophetic rings of the Brahmans had on each the name of a planet.

It is said the number seven is used more than three hundred and sixty times in the Hebrew scriptures.

The seven notes of music were used in chanting the praises of the sun.

The Egyptians had the seven mysterious vowels and the seven divine rivers.

The Phoenician Cabiri called "The Great Ones" were the special Gods of sailors and were seven, and with Esmund became eight. Esmund was the youngest son of Sydik of great beauty. Venus fell in love with him; he corresponds to the Syrian Adonis and Hebrew Joseph.

The Cabiri invented ships and presided over navigation.

That is why Jacob must serve seven years for Rachel, which was the period of the sun's exile and inferiority in winter, and is released in spring at the seventh month. And the apprentice served seven years to learn his trade. And the minor was of age at twenty-one, or the three times seven. The sick man was healed in the seventh sign or seventh month. Jericho was betrayed by the harlot, the winter moon, in the seventh month.

Beasts were taken in the ark by sevens, and the ark rested in the seventh month, which was the spring.

The Jews were carried captive in the seventh year (Jer. 524:28).

NUMBER EIGHT.

All the ancient nations acknowledged and taught the hidden properties of numbers.

The Babylonian Marduk holds in his hands the eight wands of fate; in the Egyptian the sun was anciently expressed as an eight-spoked wheel.

Eight originated at first from the four cardinal points and then increased to eight by the four intermediate ones.

The temple of Bel at Babylon was a pyramid of eight square stages.

The seven Cabiri are the seven planets and a chief like Esmund or Ptah is added to make eight, who was the sun.

The Buddhists have "the eight-fold Holy Path" and the "eight spiritual states."

The Hindoo prostrates upon the ground with eight members—two hands, two feet, two shoulders, with forehead and chest.

The Egyptian gods were divided into three or more orders, the first order of eight gods, and second order of twelve, and third the unknown.

Sacrifice on eighth day (Exodus 24:30.) Thou shalt offer the first-born; seven days shall it be with his dam and on the eighth thou shall give it me. The eighth is one the first quarter of the moon.

In dividing the month of twenty-eight days into four parts or four quarters, they become the 8th, 15th, 21st and 28th.

The Egyptians had eight great gods. These seven Cabiri of the Phoenicians were all the sons of one great father called Sydik "the just," and these eight formed the great gods of the Phoenicians and correspond to the eight gods of the Egyptians (Wilkinson, Egypt, 11:483.)

The Cabiri are the seven planets. In the Egyptian Phtha was added for the eighth. Again the same in Phoenician we have Esmund, son of Sydik the just and brother of the seven Cabiri. Esmund signifies the eighth, each seven has a patron at the head and this Ogdoad which embraced the eight primitive gods of Egypt.

The Phoenician women mourned for Adonis and sought for him seven days, and on the eighth he was found, revived in the sprouted seed of barley, lettuce, and fennel, and his phallus was drawn by oxen to the temple as the phallus of the Bull was lamented by the Babylonian women in the legend of Gilgames.

It was the personification of the new moon ring represented as the author of vegetation.

The number eight was at first the cardinal points, then four semi-cardinal points added. The gods of Egypt are four gods and their four wives that is four pairs; with temples some had four interior pillars and eight in the outer circle of columns, probably the first compass had but four points, and the measurements of land first taken from the cosmic divisions of the universe and the ground work of the temple and the augural delimitation. Among the ancient Chinese cultivated land was marked into squares and the central plot was "God's acre" and its productions were applied to religious purposes—plots had nine squares, eight secular and one in the centre sacred so that the one in the middle had two squares on each side.

Everything celestial had an earthly counterpart and vice versa. Eight was a complete number; there were eight grains and eight notes in music, eight souls were in the ark.

The King of Siam has an octagonal throne, this finally was increased to twelve, eight stone columns were placed around a temple and stones were cut with eight angles.

The Captain and crew of the celestial ship had eight souls in the ark. Sydik and his seven sons who are the Cabiri making eight.

Krishna, the Hindu redeemer, was the eighth son of Vasudeva.

The chief gods were four faced and circumcision was performed on the eighth day and the Brahman was invested with the sacred thread at eight years of age.

NUMBER NINE.

The number nine originated from the nine months of the Embryonic development or the prenatal life of man.

Consequently nine muses or sea maidens sang in the Norse creation of the world.

So the universe of the Norsemen was made to consist of nine regions, and Odin hung nine nights head downward on the tree to sweat out the ruins of wisdom.

There was a cycle of nine gods over Egypt and the Egyptians held that this cycle represented the self-development of Ra and reminds one of the nine nights Odin hung upon the tree in sacrifice "myself to myself, I offered that all wisdom I might know." They represent the nine months of the prenatal life of man—they were the nine muses or water maidens who sang around the moon cauldron at the Norse creation.

Thoth, the Egyptian moon god, was worshipped with eight cosmic forces called Sesennu, making nine.

They had a nine days fast at the initiation of the mysteries for the adoration of the two goddesses of the Eleusinian mysteries.

The Eleusinian, the great annual feast, lasted nine days, celebrated (Sept.-Oct.)

The Norse Heimdal was called the "pure white god and was the son of nine virgins who were sisters. He was the heavenly watchman and keeper of the rainbow.

And there were nine streams of Elivagar, the poison stream that flowed out of Niflheim or hell, and in northern mythology there were nine worlds and there were nine choirs of cherubim and seraphim of Chaldeans and Jews.—Dupuis, p. 54.

And we have a chief Judge and eight associate Judges making nine.

Nine was the limit of elementary numbers, there could be no elementary number beyond it. It contained all the elementary numbers within itself and was called by the Pythagoreans the horizon of numbers.

The number nine was strong among northern European nations—at the three times three. In every ninth month they renewed the ceremony of sacrifice which should last nine days and every day they offered up nine victims whether men or animals; but a still more solemn sacrifice was offered up at Upsal, in Sweden, every ninth year when the King and all the citizens of distinction were obliged to appear. (Mallett, p. 112 North Antiq.) The Scandinavian's numbered to twelve without breaking at ten. They reached twelve by units. (Mallett, p. 219.)

The Jews light the wax candles in a nine-branched candlestick. The fortifications of the Acropolis were built with nine doors, one within another.

It was anciently the custom to divide works of literature into nine books, as the genius of the writer was supposed to be inspired by the nine Muses and Herodotus is divided into nine books for the same reason.

"Hour of prayer being the ninth." Acts 3:1.

"Saw a vision about the ninth hour." Acts 10:3.

"Christ went out about the ninth hour." Matt. 20:5.

Christ cried out about the ninth hour, "Why has thou forsaken me." Matt. 27:47.

Eighteen and twenty-seven are multiples of nine and eighty-one is the square of nine and this eighty-one is the favorite number with Chinese.

NUMBER TEN.

The number ten was generally considered perfect and complete—it was the sum of the fingers on both hands, by which men were first taught to count and reckon. The old lunar year consisted of ten months, five summer and five winter in round numbers, the year of 360 days throwing off the five which left thirty-six days to the month, five summer and five winter months.

The Greek year was one of ten months, five were summer and five were winter, and the old Roman calendar contained but ten months in the time of Romulus.

The seasons of the year were represented with the Chaldeans by a genealogy of gods.

Most of the ancient political nations have the ten mythical heroes at the head of their national history.

In their traditions before the foundation of Nineveh there were ten generations of heroes, Eponyms of ten successive cities.

The Armenian traditions record the ten ancestral heroes before Aram (Lenorm "Begin. Hist." p. 29), and the Iranians have at their historical beginning ten mythical heroes ending with Gayomaretan, and the Hindus have the ten pitris or "fathers."

The Chinese have ten Emperors and so did nations who successively inherited and transcribed the beliefs of their ancestors.

Germans and Scandinavians had ten forefathers and Arabs have the ten kings at their beginning.

The Egyptians had the divine reign of ten gods and this

division of ten is very ancient and preceded the number of twelve heroes.

The ten Patriarchs of Genesis correspond to the ten solar mansions or signs of the Zodiac.

The Hebrews have ten generations from Adam to Noah, and also ten commandments, and cause ten plagues in Egypt, and have ten lost tribes.

According to Ewald, Enoch was the Noah who ended the old planetary year of seven gods and it may be at the time when solar time of 365 days superseded the old lunar year of 354 days.

And he became the solar man of 365.

That the seven or ten names originally used as solar physiognomies have been artificially reduced to a human genealogy with the names of men.

The Hebrews instead of gods turned these solar heroes into Patriarchs and concealed their astronomical character.

There were ten great virtues of Brahma, and Vishnu performed ten incarnations.

Babylon was besieged by the Sabaeans for ten years.

Thebes was besieged ten and Troy for ten years.

And the war against the Titans by Zeus lasted ten years.

NUMBER TWELVE.

In Babylonian mythology according to Diodorus there were twelve gods of the heavens and one assigned to each of the twelve signs of the Zodiac.

Twelve signs of the Zodiac.

Twelve tribes of Israel.

Twelve tribes of the Nahoreans of the north.

Twelve Titans who made war against heaven.

Twelve tribes of the Ishmaelites.

Twelve districts of the Idumeans.

Twelve children of Aeolus, the wind god, six sons and six daughters.

As one year has twelve children like Jacob, the twelve forms of Jacob the sun or year.

Twelve altars of Janus, twelve labors of Hercules, twelve Asa gods, twelve precious stones of the high priest, twelve cities of Ionia.

It is said that the rock struck by Moses issued in twelve brooks, one for each of the twelve tribes.

Twelve years service of the King of Elam (Gen. 14:4).

Twelve wells at Elim (Ex. 15:27).

The number twelve contains the five planetary gods added to the seven magnificent gods, making twelve.

The ten means the same as ten tribes of Israel, and the two chosen added make the twelve, and the two were sun and moon.

Five Paters and five Aves the Apostles Creed and the Con-
fiteor, making twelve, are to be said before leaving bed in the
morning in Catholic devotions.

Numbers 24, 36 and 48 seem to be often multiples of
twelve, as nine is a multiple of three.

NUMBER THIRTEEN.

The Norse Loke and the Hebrew Judas are called the "Thirteenth," the evil number. In Asgard were erected twelve seats for the Aesir and one for Odin, the all father.

The pyramid of Cheops covers thirteen acres.

The Aztecs of Mexico had thirteen chief deities.

A King having thirteen sons gave one to fate and the lot fell on the eldest son.

This moon is a giant with three heads.

The City of Constantiuople is surrounded by a wall about thirteen miles in circumference and has twenty-eight gates open upon the city. The thirteen represents the thirteen lunar months of the year and the twenty-eight the days of a lunar month.

NUMBER FOURTEEN.

Amphion (the sun) who built the walls of Thebes married Niobe, the moon, and had seven sons and seven daughters who were slain by Apollo and Diana.

They were visibly the moon rings or months for Apollo and Diana became destroying divinities at winter.

They are the seven young men and seven women sent every year to Crete to be devoured by the Minotaur or moon Bull, and the seven lean and fat of Pharaoh's time.

They are the fourteen pieces in which Osiris, the Egyptian god, is torn every year, and one ring escapes.

And the fourteen children of Niobe are the fourteen days of her increase, and the death of her children are the fourteen days of her decrease—which make the twenty-eight days of her life until dead, when a new moon appears.

The tomb of Osiris as it has been excavated has the shape of a dwelling containing fourteen rooms, as he was torn while living into fourteen pieces. There are five rooms to the north, five to the south and four to the east and the western face is open, the place of sunset where he disappears.

That moon is the box in which Osiris the Egyptian Christ was placed by his evil brother and sent to Hades—(Pandora's box) and drifted away to Byblos, the book or record city or wisdom city and lodged in a tree and the limbs of the tree were stripped away and left a pillar in which was the soul of Osiris, and his wife Isis followed weeping like the mother of Christ and begged the pillar, and this story of the god Osiris went back almost to the beginning of Egyptian civilization.   The above tree was the moon of fourteen limbs or rings.

The Egyptian Sphinx is hewn out of a living rock, the body 150 feet long, the paws 50 feet long, the head 50 feet long, the face 14 feet wide and from the top of the head to the base of the monument the distance is about 70 feet.   All these represent sacred and conventional numbers.   The Sphinx was the moon.

NUMBER TWENTY-ONE.

The sun entered Aries, the Ram, on the twenty-first of March, the beginning of spring and the renewal.

Twenty-one was a very sacred number with the ancient Persians. Their most sacred prayer the Honovar, contains twenty-one words, and the original Avesta contained twenty-one books, "Zend Avesta Britannica;" and in law, a minor is one who is under twenty-one years.

Rig. Veda, vol. 5, p. 125, "When Indra and I ascend to our home, the world of the sun, then having drunk the sweet soma, let us be united in the 21st sphere of the universal friend."

Ulysses had been absent twenty years when he arrived at home, conveyed in a Phaeacian vessel to the shores of Ithaca. No man could be kept in outlawry more than twenty years.

Twenty-one guns are fired for a salute or three times seven.

The City of Rome was marked out by a plow on the twenty-first of April.

Jason, the one sandaled man said, in reply, "I have come from the cave of Chiron (the winter moon)." I have accomplished twenty years; I have come to claim the Kingdom of my father.

The thrice seven on the twenty-first day was the birth feast.

NUMBER FORTY.

Forty was a gloomy number, one of trial and suffering.

The Babylonian Gilgames, the sun god, was forty days on his winter journey, being ferried over the waters of death to the Isles of the Blest. It was the time of Noah's flood, when it rained forty days and forty nights (Gen. 8:12).

Lent was the period of sorrow when the cluster of the seven stars in the constellation of Taurus the Bull disappear for about forty days, they were called the ship stars by which mariners kept their course before the discovery of the compass.

Hesiod says the Pleiades or the ship stars, which are seen in the neck of Taurus, were the daughters of Atlas and there was a time in the spring when they were absent forty days or were hidden by daylight, but after that time return to the night sky, and the Egyptians found their sacred Nile to be about forty-two days in rising (six weeks) and the sun was forty-two days obscured. The Pleiades, the great ship stars of the mariner, were observed for about the same time and they divided Egypt into forty-two districts and had forty-two commandments.

The winter moon became the den of thieves who opened the moon door by a password—saying, "Open Sesame."

It was the den of the "Forty Thieves," and the "Robbers Roost," and where "Moses and Christ Fasted forty Days," and the cave of the "Seven Sleepers," and the winter cave of "Hamlin the Piper," and the Nile was forty days in rising, which was the great hope of Egypt, and forty days were set apart for fasting and prayer to propitiate the Nile god.

This was the season of the sun's leanness in his dormant state of winter or Lent.

Fasts were prolonged forty days, especially the vernal fast in spring.

It was the forty days when the ruling star is in evil aspect according to the astrolabe, and the Etesian winds prevailed for forty days after the setting of the dog star.

The Druids and Mexicans celebrated the death of the year at the midnight culmination of the Pleiades.

The Aztecs at the termination of the great cycle of fifty-two years held a great festival, put out fires and the new fire was kindled upon a mountain by friction at midnight as the constellation of the Pleiades approached the Zenith.

Saul, David and Solomon reign 40 years.

Gideon and Deborah judge Israel 40 years.

Eli judged Israel 40 years.

The Philistines reigned over Israel 40 years. (Judges 13:1).

After Othniel, the Judge, delivered Israel from Cushan, King of Mesopotaniia, the Israelites rested 40 years (Judges) —under the reign of Ehud, son of Gera, the land of Israel rested 40 years (Judges).

Again in the time of Deborah there was rest 40 years.

Israel delivered into the hands of the Moabites 40 years (Judges 13:1).

Goliath defied the armies of Israel 40 days.

Elijah fasted 40 days.

Forty days were fulfilled for Jacob, for so are fulfilled the days of those who are embalmed (Gen. 50:3).

There fell of the Ephraimites forty-two thousand (Judges 12:5).

There were forty-two children Elisha cursed to be devoured by two she bears (11 Kings 2:24).

The principal works of the Jews, according to the Talmud, were forty less one to escape the evil number forty.

Thirty-nine is forty, save one, the thirty-nine books of the Old Testament are purposely one short of the evil number forty.

Hengist, the Saxon invader of England, reigned 39 years.

As Jews divided Canaan into twelve parts and their people in twelve tribes to agree with the twelve months of the solar year, and regeneration was taught by the renewal of the year.

Forty was a very common number for pillars and walls—properly the original number is forty-two or six weeks, but the round number of forty is substituted for convenience.

Forty pillars were used in the construction of temples.

In Persia with the temple of Chilminar, the same with Baalbec which had forty pillars, also in Druidical temples as at Stonehenge. In the Hebrew temple the doorway of the porch was forty cubits high and twenty broad and over it were five beams of carved oak.

Barclay's Talmud, p. 262.—"And the Israelitish spies were sent to spy out the land of Canaan and searched the land forty days."

The outermost wall of Babylon was forty-two miles in circumference or 360 stades.

The golden image of Bel at the temple of Babylon was forty feet high and there was a golden table there forty feet long and fifteen broad.

Moses was forty years old when he fled from Egypt, and was forty years with Jethro tending his flocks.

And Moses abode in Mt. Sinai forty days and forty nights, waiting for the ten commandments of the law, eating no meat and drinking no water. (Exodus 24:9—18.)

Jonah 3:4, "yet forty days and Nineveh shall be overthrown, a city of three days' journey."

It takes the messenger of the sun three day always to arrive at the moon; it is on the third day of the moon's darkness the angel of light, the solar preacher and priest of the altar, arrives at the moon.

Isaac married Rebekah, daughter of Bethnel, in the fortieth year of his age.

And Esau at forty married Judith, daughter of Beeri the Hittite.

Advent formerly occupied six weeks or forty-two days in honor of the approach of the anniversary of Christ's birth, and the festival of the 25th of December. The birthday of Mithras was kept for forty days.

And the Egyptian god Osiris was said to be dead or absent forty days in each year, during which he was mourned, corresponding to the exile of Moses and Christ.

And the Syrians did the same for Adonis and Scandinavians for Frey.

And the limbs of Osiris were scattered and sought for forty days and so were the bones of Bacchus in Gaul.

Forty Egyptian judges came and seated themselves in a semi-circle (like the half moon altar) above the lake where the dead body is to pass over in a boat called Baris in charge of a pilot called Charon, and in Egyptian, Kare (Diodorus, 1:92.)

The chambers of the Egyptian serapeum containing the mummies of the sacred Bulls are forty in number.

The sacred books of the Egyptians were forty-two in number, edited by Thoth, the inventor of writing—he was the moon corresponding to the Greek, Hermes.

These Hermetic books of the Egyptians ascribed to Hermes contained the sum of all knowledge both human and divine.

In the Judgment hall of Osiris forty-two judges passed sentence on the forty-two sins of the deceased person and there were forty-two mortal sins.

At the end of every twenty-five years whenever a new bull, the Apis, was found, the calf was taken to Nilopolis and there the Egyptian women stood before the bull for forty days (during which women only were allowed), by which they secured the blessing of a numerous progeny.

At the summer solstice the sun appears to rise at the same point of the horizon for three days together and at that time the greatest amplitude of the sun is forty degrees at rising and the same in the west at setting.

And the greatest ampiltude at rising in winter is 40 degrees south of east and the same amplitude at setting south of west.

The Serapion or temple of Osiris which contains the mummies of the sacred bulls each of which is placed in a chamber, and the chambers are forty in number and excavated on each side of galleries about twelve feet wide, and on one side of the labyrinth close to Lake Moeris stood a pyramid forty feet high.

The Egyptians observed forty days for those who were embalmed.

Abury works are constructed upon a huge serpent whose bow is curved to the north, extending for miles. The head of the serpent is composed of forty stones enclosing a temple of nineteen stones referring to the moon cycle.

On the summit of Babel was the shrine in which stood the golden image of Baal, 40 feet high, and a golden table forty feet long and fifteen broad. This Babel was a pyramid of eight square stages.—Babylon "Britannica."

It rained forty days in the time of the flood and so the great reservoir was constructed forty miles square on the west of Borsippa which receive the waters of the Euphrates.

The temple of Heliopolis was surrounded by forty pillars and these magical astronomical numbers were common to all the principal religions.

Forty conspirators landed to slay Paul. Acts 23:13.

Christ fasted forty days.

Forty stripes given in chastisement. Deuteron., 25:3.

The beast to which was given to utter blasphemies for forty-two months. Rev. 13:5.

And in the religious festival of the Persians the salutation of Mithras continued forty days.

In the twelfth century a church penance was given for forty days on bread and water.

In the graveyard of the Swiss Lake dwellers along with each skeleton were found forty flakes of the tusks of the wild boar pierced at each extremity, which shows they had the same Adonis worship of Babylonia and Phoenicia.

As a burial custom in ancient Scythia the dead Scythian was carried about among all the neighbors for forty days and a funeral feast was given by every friend so visited.

Maidens and many horses were sacrificed to the departed spirit of Jenghix Kahn by his successor to the throne in the 13th century and again at the election of Kuluk Kahn. During the feast, which lasted a week, forty oxen were consumed each day. ("Mongols" Ency. Brit., p. 742).

NUMBER FIFTY.

Fifty-two seems to be the cycle of thirteen lunar months of the year multiplied by the sacred four equals fifty-two, the lunar weeks of the year.

Chaldeans had seven supreme gods in the seven planets and fifty great gods in the fifty-two weeks of the year in round numbers.

Giants were fifty headed and one hundred handed.

The Hydra had by some one hundred heads, by others fifty.

Cerberus, according to Hesiod, had fifty heads and the sea god Nereus had fifty daughters.

The Nereids or nymphs of the sea were said to be fifty in number.

Endymion Danaus and Priam Aegyptus and Lycaon all had fifty sons or daughters.

NUMBER SEVENTY.

The number 70 was established by the precession of the equinoxes. The sun advances about one degree in seventy-two years causing the spring equinox to occur one degree earlier every seventy-two years or advancing about one sign in 2160 years and completing the entire circle of the ecliptic in about 26,000 years which was called the great year; hence the sacred seventy in round numbers.

The precession of the equinoxes is caused by the reeling motion of the earth which is caused by disturbing attractions, hence the seventy elders of the Sanhedrim and the Yacna of the Persian Avesta contained seventy-two chapters and the girdle of the Parsee contained seventy-two threads.

Typhon was in league with 72 comrades to slay Osiris, his Egyptian brother, and the Egyptians mourned for Jacob seventy days. Gen. 50:3.

At the funeral of an Egyptian King, according to Diodorus, the temples were closed and the people held no sacrifice or festival for 72 days.

Sargon, the King of Babylonia, caused the great work on astrology to be compiled in seventy-two books which were translated in Greek by Berosus.

Ex. 1:15—"And all the souls that came out of the loins of Jacob were seventy."

Ahab had seventy sons in Samaria.—II Kings 10:1.

Jeremiah (25:11) served Babylon seventy years.

Luke 10:1—The Lord appointed seventy.

NUMBER 100.

The Zodiacal signs were the arbitrary divisions of time for convenience and the 12 signs each had a spirit which presided over every month, and the human body was divided into twelve parts, each part of which was under the influence of that spirit when passing through that sign.   Circles had 12 pillars as 12 stones set up in the Jordan.

In ancient times of the formation of the Zodiac, some had twelve signs and others but ten.

The great hundred was ten twelves, 120, called "the long hundred," and was a duodecimal computation formerly common in Britain and Scandinavia.

A staff of reeds was the omen stick of the British Druids, and we still retain the word Hand-reed contracted as heard in our numeral, one hundred, a contraction of Hand reed.

Babylon and Thebes had each 100 gates.

The Sleeping Beauty and Barbarossa were condemned to sleep 100 years duration or the one hundred days of winter.

Hercules delivered his country from the annual tribute of 100 oxen which it paid to Erginus.

Gen. 33:19—And Jacob bought of the children of Hamor a parcel of a field for an hundred pieces of money and erected there an altar and called it "El-Elohe—Israel God"—the God of Israel.

The Nacissus had 100 roots.

The winter dragon that guarded the apples in the Hesperian Garden had 100 heads and never slept. All these are the hundred days of winter.

Argos was hundred-eyed, Hydra one hundred headed and Briareus had one hundred hands.

And there were a hundred men concealed in the wooden horse presented to Troy and the horse was dedicated to Athene by the departing Greeks.

All the above represent the period of winter.

Then the solar hero of ancient Britain says of himself, "Have I not destroyed a hundred forts, have I not slain a hundred governors, have I not given a hundred veils and slaughtered a hundred chieftains?"—Davies' Mythology of British Druids, p. 560.

The narcissus plucked by Prosperine had 100 flowers which grew from one root.

The Hecatomboia or one hundred oxen sacrificed to Juno and the flesh distributed among the poor. Again the Hecatomphonia, a sacrifice to Jupiter when a hundred enemies had been slain.

Hecatompolis was named for its hundred cities.

Hecatompylos Thebes for its 100 gates.

The children of Israel offered a hundred bullocks at the dedication of the House of God. Ezra, chap. 6, verse 17.

Minos paid a vow to Jupiter, the bodies of a hundred bulls, after overcoming the Athenians, when he reached the land of the Curetes (Metamorphoses p. 269). The eyes of Argus were 100, the same as the heads of Hydra which reach a hundred (Ovid, Metamor. p. 303). They are not stars nor those of Briareus or Cacus, they are definite cycle numbers which refer to the period of winter generally called the one hundred days.

Again the maid stung by a spindle who fell in a sleep of 100 years duration, and at the end of that time awoke from her sleep by the kiss of her hero lover, the kiss of the spring sun.

Barbarossa has to sleep 100 years, when the winter ravens will cease flying, then he will break his chains, conquer his enemies and hang his shield on a withered pear tree which will again blossom.—(Asgard and the Gods, p. 80).

In large numbers the 100 is most common and the thousand and one and the year and a day.

Kalevala, Vol. I, p. 19—Wainamoinen planted the first oak and it grew evil and with its branches hid the sun, and he employed a pigmy from the copper kingdom to cut it down, an oak of a hundred branches, and in its fall it was shivered in pieces and whoever obtained a piece became master of magic and eternal welfare. The oak is the winter moon which is cut in pieces by the young stripling, the new moon ring, all its branches and twigs are healing in spring—the pigmy who cut it down is the first rod or ring of the spring moon which is the axe that hews down the black winter moon, a limb every night.

Iphidimas gave 100 oxen as a wedding present to his father-in-law. (Iliad XI, 244.)

There are 100 bright moon rings or three months of winter.

Ptah-Nefer-ka, the Egyptian god, gave one hundred pieces of silver for the hidden book of wisdom which was the moon, the 100 oxen the father of Odyssus gave for the old nurse, the servitude of the 100 days of winter exile.

The Hydra is the serpent of a hundred heads. Like Argus, he is the winter moon, 100 days of winter in round numbers.

Hercules defeated and killed Erginus, the King of Orchomenos, and delivered the Thebans from the annual tribute of one hundred oxen. The above story is transparent— Hercules, the sun, killed Orchomenos, the winter king of Orcus or Hades, the moon king, and delivered the summer race from the annual winter captivity.

MISCELLANEOUS NUMBERS.

There were 360 degrees of a circle and there were 360 eons or Genii and there were 360 days of the year in round numbers and 360 idols of Japan in the palace of Dairi.

360 statues surrounded Hobal, the sun god Bel of the Arabs, and 360 chapels built round the mosque of Balk.

And the Orphic theology recognized 360 gods.

This solar wheel sometimes said to have 360 spokes in the Rig. Veda, and in the three axles of the wheel, three seasons.

The Egyptian priests daily poured 360 cups of water to the Nile, the number of days in the year.

We are told of the three hundred warriors of Thermopylae and the three hundred warriors who lapped water as a dog under Gideon's command (Judges 7:5), and the three hundred of Abraham's household.

300, 600 and 6,000 might be multiples of 60.

300 is the number of days in the Roman year of ten months.

30 is the number of days in a month.

1000 or ten thousand are indefinite numbers—that is a number of a great multitude, as in the Hindu Puravas the solar orb is called the thousand rayed.

Indra, the god of the firmament, called thousand eyed.

Krishna has ten thousand wives.

The Phoenix rose from its ashes once in five hundred years.